BUILT ON SAND

BUILT ON SAND

THE SCIENCE OF GRANULAR MATERIALS

ETIENNE GUYON, JEAN-YVES DELENNE,
AND FARHANG RADJAI

TRANSLATED BY ERIK BUTLER

FOREWORD BY KEN KAMRIN

THE MIT PRESS   CAMBRIDGE, MASSACHUSETTS   LONDON, ENGLAND

This book was set in ITC Stone and Avenir by Toppan Best-set Premedia Limited. Printed and bound in the United States of America.

Library of Congress Cataloging-in-Publication Data

Names: Guyon, Etienne, author. | Delenne, Jean-Yves, author. | Radjaï, Farhang, author. | Butler, Erik, 1971- translator. | Kamrin, Ken, writer of foreword.
Title: Built on sand : the science of granular materials / Etienne Guyon, Jean-Yves Delenne, and Farhang Radjai ; translated by Erik Butler ; foreword by Ken Kamrin.
Other titles: Matière en grains. English
Description: Cambridge, Massachusetts : The MIT Press, [2020] | Originally published as: Matière en grains, by Odile Jacob, 2017. | Includes bibliographical references and index.
Identifiers: LCCN 2019025843 | ISBN 9780262043700 (hardcover)
Subjects: LCSH: Granular materials.
Classification: LCC TA418.78 .G8513 2020 | DDC 620.1--dc23
LC record available at https://lccn.loc.gov/2019025843

10 9 8 7 6 5 4 3 2 1

CONTENTS

FOREWORD

There is something deeply intriguing about a pile of grains. On its face, it seems simple. What could be so challenging about a packing of many objects? And yet, the history of granular materials science begs to differ. Despite the commonality of granular media in day-to-day life—including many important geophysical and industrial applications—the modeling of this material displays curious challenges beyond what is seen in other common materials like water or structural solids. To understand the behavior of a landslide, for example, is to understand the culmination of physical processes happening all the way down to the scale of the individual grains and their contact and dynamic laws. These curiosities are what drew me to the field as a student and have kept me interested for years.

Luckily, the last two decades or so have seen a growing revolution in granular media, guided by rapid computer modeling techniques, innovative experimental methods, and new physical insights that span the grain level up to the geological scale. Our ability to translate particle-level information, such as surface roughness, shape, size distribution, and wetness into large-scale models and simulations is at an inflection point. Indeed, it is now timely for a book like this to help us put all these pieces together.

The three authors are leaders in this modern wave of the field. I have had the pleasure of extensive interactions with Dr. Delenne and Dr. Radjai, in the United States and in France. We organized an international conference in 2014 in Montpellier and Dr. Radjai spent several years as a visiting researcher at MIT. Over the last half-decade, we have had numerous exchanges and served on committees for each other's students and post-doctoral scholars. My appreciation of them as friends is surpassed only by my respect for them as scientists. Delenne and Radjai have developed some of the best computational tools out there for simulating granular materials—be the grains jagged, wet, soft, or bouncy. Their methods are used sometimes in place of experiments due to their reliability and have played a significant part in the development of novel theories of granular mechanics. As for Dr. Guyon, his seminal work in hydrodynamics and granular media was fundamental in my own education. But in synthesis with his technical work, Dr. Guyon has worked intensely for improving scientific communication to the public. I was fortunate to get to see Dr. Guyon present on the history and creation of the French museum system and its successes in public outreach. His drive and passion to communicate the joy of science, regardless of the audience's background, is evident in the presentation of this book.

This book is not just a work about the science and applications of granular media. It is a history book as well. It is also a how-to book of sorts, explaining methods for how these materials have been made and manipulated since early civilization. The discussions are broad, ranging from ancient ways to shape flint, to the science of making couscous, to the statistics of percolating chains, to the etymology of our words. Such breadth does justice to a subject matter rife with so many different directions to study. While the common link is granular media, this text is really an opportunity to witness a storyline that passes through chemistry, physics, astronomy, math, and engineering.

Enjoy the ride!

Prof. Ken Kamrin
Department of Mechanical Engineering
Massachusetts Institute of Technology

PREFACE

Besides water, materials composed of vast amounts of tiny grains constitute the most abundant form of matter present on Earth. Many fields of study have—often independently—developed analytical concepts and tools for describing their polyvalent properties, whether liquid or solid-like. Since the classical age in Greece, scientific reflection on condensed matter (solid and liquid) and dilute matter (gas and molecular matter) has drawn inspiration from the example of assemblies of grains. At the beginning of the Age of Enlightenment, Charles Augustin Coulomb devoted attention to avalanches moving down an embankment. About a century later, Osborne Reynolds marveled at the complexity caused by friction and geometric congestion between grains. However, only in the 1960s were the first models developed that could be applied to soil engineering. At the same time, simple models for grain flow were pioneered in fields of great economic value—for instance, process engineering, agronomy, and pharmaceuticals.

The 1980s witnessed efforts to bring the research conducted in various sectors into a unified science of granular media. Researchers modeled granular matter by means of a "bag of marbles," a model that abounds in paradoxes and differs strikingly from the picture afforded by classical solid-state physics. This model has come to represent a veritable prototype, the paradigm for granular media: both the basis for simple table-top experiments and the foundation for high-level collaboration and research.

An earlier book, *Du sac de billes au tas de sable* (*From a Bag of Marbles to a Pile of Sand*, 1994) by Etienne Guyon and Jean Paul Troadec discussed these advances and research programs up to 1990, especially in France. Together with Nobel laureate Pierre Gilles de Gennes, Etienne Guyon established an open community known as MIAM (*Milieux Aléatoires Macroscopiques*; [Macroscopic Random Media]), which marked the starting point for intensive research on an international scale for over twenty-five years. Today, it is no longer surprising to find a sandcastle on the cover of a serious journal such as *Nature,* or to see the stability of its structure being studied as a function of water content at the European Synchotron Radiation Facility in Grenoble or on zero-gravity flights. Important applications are already evident in the technological revolution of high-performance concretes—for example, large-scale building projects that capitalize on an improved understanding of the structure of materials at the granular level.

If experimental models still play a vital role in research, the intensive use of numerical simulations represents a new development. From the 1960s to the 1980s, computers started being used to describe dense, disordered phases of matter at the atomic level—liquids and glasses. The methods pioneered in this context bore fruit for the study of macroscopic granular systems. In turn, this opened the way for studying both the forces acting among the grains themselves and the rich phenomenology of their collective motion.

This book stems from collaboration between two researchers working on new developments (Jean-Yves Delenne and Farhang Radjai) and the author (Etienne Guyon) of the aforementioned title. The work at hand introduces new fields of application—in particular, soil science and agronomy—and seeks to satisfy the curiosity of a broad, and extremely diverse, readership. It surveys recent research on granular media, especially by means of computer simulations in two and three dimensions. New developments in imaging have enabled scientists to understand organized ensembles of macroscopic grains; such metamaterial presents a wide array of geometries and compositions with novel optical, mechanical, and acoustic properties suited for a host of uses. The pages that follow have also benefited from contributions presented at the World Congress, "Powders and Grains" (in Montpellier), organized by Radjai and Delenne in 2017.

This book is addressed to a broad scientific community interested in a form of matter that defies the standard scheme of classification in terms of solid, liquid, and gas phases. But granular materials can also provide a model for other domains, which lay readers will recognize. Disorder, bulk, and grain flow bear on the social and "human" sciences, as well. As our attention shifts from grains to piles, one order of reality yields to another at a higher level, illustrating the philosophical mode of argument known as *sorites*. In this light, the properties of granular media serve to describe collective behavior in disordered settings in general.

We begin by describing the single grain (whether mineral or vegetal) and accounting for its varied manifestations. Chapter 2 explores how grains are made and prepared for application; the process is extremely costly in terms of the energy expended. In chapter 3, we examine grains in piled or "stacked" form; illustrations of geometrical models provide a point of reference. Chapter 4 explores the packing fraction of granular media, a crucial issue that bears on the properties they display in use. In turn, chapter 5 examines small-scale deformations in piles of disordered grains caused by external factors, with particular attention to friction. Having established the properties of granular matter, chapter 6 introduces various modes of disorder in general terms. In chapter 7, a discussion of force chains enhances our understanding of granular media when at rest and subjected to a constraint. Chapter 8 introduces a new, dynamic model to account for collisions and friction between grains. Chapter 9 expands the view by introducing a fluid to the equation; a sandcastle exemplifies the cohesion created by the capillary action of water. Lasting cohesion (which is not a feature of sandcastles!) occurs when grains in very different states of consolidation are assembled (chapter 10). Since a fluid can also flow between grains, chapter 11 examines conditions in a porous medium, where grains retain a fixed position on the whole. The last chapter describes granular flow in a fluid. Following a simple account of fluid effects on the movement of grains, we explore ways to take friction and collisions between grains into account. This discussion expands the model presented in chapter 8, which did not consider the role played by fluids. This final, schematic portion opens up an immense variety of situations encountered both in nature and in applied settings, the province of specialized studies.

1

GRAINS, SEEDS, AND POWDERS

To see a World in a Grain of Sand
And a Heaven in a Wild Flower,
Hold Infinity in the palm of your hand
And Eternity in an hour.
William Blake, "Auguries of innocence"

Highly diverse in shape and origin, grains of matter occupy a key position in our environment. Granular matter, our object of study in this work, represents the result of a vast number of grains combining in a pile, just as a crowd represents many people coming together to form a group. The phenomenon we will be examining in the following pages is a physical reality, and countless examples abound all around us, on a human scale. This opening chapter is devoted, above all, to the basic unit of matter, the tiny object that, in analogy to the seed of plants (*granum* in Latin), is called a grain.

A CLOSE LOOK

MANY POWERS OF TEN

To understand granular matter as a whole, we'll start by examining a single grain. The field to be explored is vast indeed. All states of matter admit description as an assemblage of particles of highly variable dimensions, even though, in empirical terms, certain properties of organization recur on very different scales: the atoms of a crystal stack up just like a regular pile of oranges! At the most fundamental level, we have atoms—the elementary constituents of matter. Nuclear physicists view atoms as a collection of *elementary particles*. At the other end of the spectrum, a meteorite may serve as an example of a very large grain. The difference between a grain of so-called *colloids**[1] (for instance, the particles suspended in milk that give it a whitish color), a powder that weighs less than a millionth of a gram, and a meteorite involves more than twenty-four powers of ten!

In this book, we'll be limiting ourselves to grains of intermediate size; hereby—as the boxed text that follows explains—*thermal agitation** generally proves to be negligible. As we will see, this is the case for powders and sand on the beach, just a hundred micrometers in diameter. It can be difficult to appreciate that, after water, grains like this represent the most frequently used material on our planet.

In the course of this chapter, we will define what scientists mean by a "grain," which, together with many other grains, makes a "pile" or a "packing." This initial characterization will be retained throughout the book. Here, the only properties of a single grain to receive our attention are those relevant to later chapters.

Of course, if we were discussing how to use these grains in chemical reactions, or how to account for their mechanical or electrical properties, further distinctions would be necessary. This information will be introduced at a relevant juncture, in keeping with the subject explored. For the moment, however, a geometrical description of the individual grain will suffice.

TINY GRAINS

A crystal may be viewed as an ensemble of atomic grains that are well ordered and fixed in place. A liquid also consists of atoms or molecules in a compact state, but they are disordered and mobile. Finally, a gas comprises the same elements in dispersed and diluted form; they can be pictured as a mass of *nanoscopic** bumper cars. Grains just a few angstroms ($1 \text{ Å} = 10^{-10}$ m) in diameter are constantly colliding—a manifestation of the gas's absolute temperature. Such movement can overlap with mean

1 Words and terms followed by an asterisk are explained in the glossary at the end of the book.

continuous flows on a larger scale, which belong to the realm of fluid mechanics.

At a scale slightly larger than that of atoms, one finds small particles, or micrograins, dispersed in liquid. One example is the aforementioned microdroplets suspended in milk that diffract light and make it look whitish. These particles retain a certain thermal agitation that decreases as their mass increases. A continuum extends from objects the size of a molecule to those the size of these colloids, which are ten to a thousand times larger than an atom; thermal agitation declines in proportion to growing size. Programs of research and applications in this domain—which is that of the nanosciences—have recently charted major developments on the basis of new possibilities of preparation and manipulation; their study must be reserved for a separate work. Here, we will generally limit ourselves to objects bigger than a few microns, for which thermal motion is insignificant. The objects in question are grains large enough to be felt "in the palm of your hand."

Thermal Agitation

In 1827, the botanist Robert Brown first described the continuous and disordered movement exhibited by small grains of pollen suspended in water, thereby renewing an observation made by Lucretius some two thousand years earlier! This motion is a manifestation of thermal agitation affecting inert particles of matter in permanent collision with each other. The root mean square of these particles' velocity v_T is related to their mass m and to absolute temperature T by $\frac{1}{2}mv_T^2 \approx k_BT$, where k_B is the Boltzmann constant (which is universal) and "$\approx$" means "of the order of." When applied to the molecules of a gas, the formula yields a velocity of several hundred meters per second, depending on the value of m. This velocity is visible under a microscope for colloidal particles a few thousand angstroms in diameter, with a mass a million times larger. In contrast, it is completely negligible for grains of matter larger than a few microns (which we mainly will be discussing in this book). The thermal agitation of particles that settle in a fluid limits their accumulation: on the surface of the Moon, for instance, it is sufficiently large that gravity cannot hold oxygen near the surface. In general, the density of Brownian particles decreases rapidly (exponentially) the higher the altitude and the weaker the gravity. Using this law when conducting research based on Albert Einstein's theoretical work, Jean Perrin (figure 1.1) was able to determine the Avogadro constant N_A (the number of gas molecules contained in a volume of 22.4 liters). He received the Nobel Prize for the experiment.

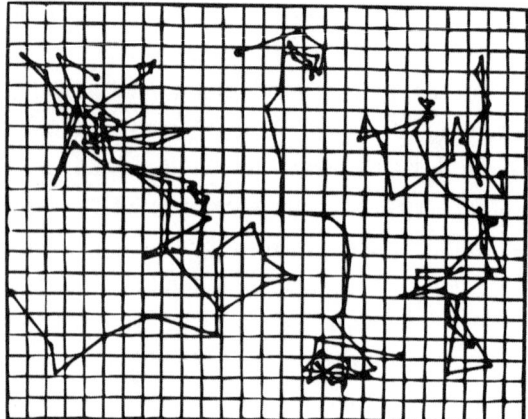

Figure 1.1
A one-micrometer mastic particle's random motion observed under a microscope by Jean Perrin.

The jagged lines on figure 1.1 might have been produced by the random motion of bacteria propelled by their flagella. In such a case, it does not make much sense to define an equivalent temperature for characterizing the agitation of "biological objects" that are *active* (in the sense of the bumper cars mentioned earlier). This book considers only *passive* objects without an internal motor. Like pucks in a game of air hockey, macroscopic grains can remain in a constant state of agitation if they are constantly shaken; all the same, they're still passive objects!

LARGE GRAINS

We have eliminated excessively small grains from our field of study. At the other end of the spectrum, are there objects too large to examine? Does a planetary object count as a grain? The rings of Saturn, which are readily visible through a small telescope, are composed of a very thin layer of ice particles and dust the size of what we find on Earth. Determining an outer limit is tricky. The meteorites—"space stones"—that reach our planet have sizes, shapes, and chemical compositions that might be mistaken for those of "Earth stones." Is this so surprising, given that our rocks have resulted from extended periods of disintegration and reconstitution here on Earth? Indeed, our planet itself emerged as an agglomeration of meteorites made up of grains with different chemical constitutions, perhaps in the wake of a supernova's explosion. Certain elements (gold and heavy metals, for instance) probably have such an origin; at any rate, they could not have been synthesized in the stars themselves. The elements stuck to each other and separated through sedimentation, with heavy ones like iron gathering at the core and light silicates forming at the crust. Although this process has slowed down significantly, several tons of these grains fall to Earth every year. The smallest aren't more than a gram; but the huge rock that made Meteor Crater in Winslow, Arizona, weighed roughly three hundred thousand tons! Today, the physics of granular media also concerns planetologists; we will return to this point.

GRAINS THAT CAN BE HELD IN THE PALM OF ONE'S HAND

William Blake invites the reader to contemplate a whole "world in a grain of sand." Following these inspiring words, we will now look at *one* grain, just big enough to be felt, whose shape and surface state may be readily discerned through a magnifying glass or microscope. The "infinity" that opens up is the voyage over the Earth that a single grain can make across geological time up to the present day. Our grain, which is too large to be subject to thermal motion, reacts to gravity and external forces. Every step we take leads through history, and we tread transformations of the mineral and animal worlds underfoot. Without even realizing it, we're kicking up the documents and archives housed on the Earth's surface.

A professor and his students—grain collectors, or *arenophiles*—created a museum of sand on at a sea resort in Château d'Olonne, *Planète sable* (*Sand Planet*). There, for universal admiration, the rich array of samples they collected or were given is on display; the young curators share their aesthetic sensibility with visitors by means of lenses, which activate the "memory" stored in the grains (figure 1.2). Taking a close-up look resembles the way a geologist observes a stone when performing research: seeking clues, magnifying glass in hand. A Sherlock Holmes of petrography, the late Maurice Mattauer was a professor at the University of Montpellier and author of a remarkable book, *Ce que disent les pierres* (*What Stones Have to Say*). There was nothing like going on one of his geological excursions through the Cevennes mountain range he loved. Just clean

Figure 1.2
Like many other examples one might choose, this sand, gathered on a beach on Crete, illustrates the diversity of sizes, shapes, colors, and even base material. Planète Sable Collection, Château d'Olonne, France.

the surface of a pebble, feel its heft, and turn it around (without much attention to the grains immediately surrounding it—it's been traveling for so long ...), and a whole history, genealogy, and vanished epoch came back to life. A little spit and polish is all it takes to attenuate the surface imperfections diffusing the light, and all the contrasts and internal patterns of unprepossessing stones emerge. Now, the merest pebble is fit for a jeweler's display case. One discerns material structure due to heterogeneous composition as well as the *deformation** experienced as resulting from a tectonic journey: stripes, furrows, and deposits. As we proceed, we shouldn't forget the humble pebbles on the way, which also have a story to tell (figure 1.3a).

We can take the same approach to a single grain. Which one to choose from a handful gathered on the beach, which all have different properties and points of origin? What's the common feature shared by that slightly translucid grain of white sand and the one next to it, entirely black, both of which may have traveled thousands of miles to wind up here? The first one is quartz, perhaps the result of *sandstone** breaking down. It's more

Figure 1.3
The varying shapes of these grains tell us about their past: a) stones of various but rounded shapes in a river; and b) excavation of highly angular stones produced as the result of *fracture** by frost. The geological hammer in the picture provides an indication of scale.

angular and might have come from a crystal inside a granitic rock. A grain of quartz is so hard that the back-and-forth of particles rubbing together as ocean waves move barely softens its forms. Yet again, the shape bears a record of its origins. The black particle, which is denser, might be a fragment of basaltic lava—the kind that abounds on the beaches of volcanic islands like Hawaii. This kind of grain is quite different and humidity breaks it down rapidly.

We can also find colored grains, very bright ones, whose irregular shape directly reveals their origin: they're elements of broken, worn, and carbonated shells with a chemistry that has partially dissolved in the sea (which is already rich in carbonates). A grain like this may also represent the result of a rock that cracked because of ice or lightning. The protracted flow of a glacier may have chiseled it and ground it down (figure 1.3b). Successive shocks might have fragmented it before it was carried in a river or borne through the air in a process called "saltation," causing it to land at various locations before settling in a dune; sometimes, it will have traveled at an altitude of thousands of meters over thousands of kilometers—like the sands of the Sahara that make scattered red stains on Nordic glaciers. Perhaps this grain of sand has been "recycled"; originally part of a rock that broke apart, then subjected to abrasion by air and water, it may have sedimented once already and formed part of a block of sandstone, only to be sent on its way again before reaching us after a "second birth." This "geological cycle" represents the very motor of changes to the Earth's surface (figure 1.4). Investigation proceeds by looking at the marks that record the grain's travels; if the stone is big enough, the deformations will be plain.

This history permits one to draw up an "identity card" for grains. For quite some time, it was standard practice among soil engineers to use the geological origins of grains when referring to them. This science—*pedology**—was long the province of naturalists and chemists like Justus

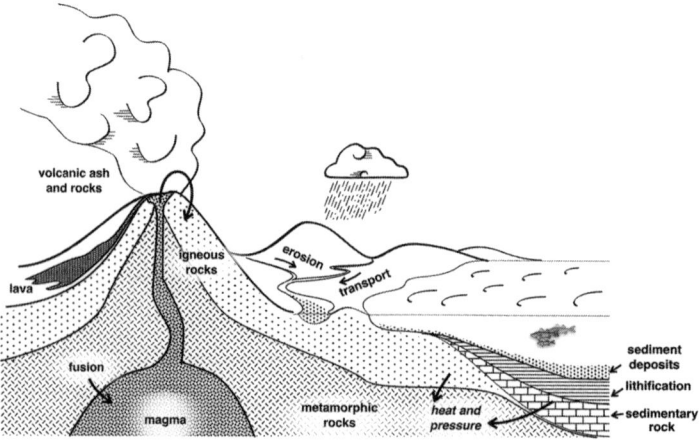

Figure 1.4
This illustration provides a schematic representation of the possible cycles in the long "life" of a grain.

von Liebig (who is better known, today, for having invented a certain bouillon cube, or for having quarreled with Louis Pasteur). Only in the course of the twentieth century were methods devised for characterizing grains in mechanical and physical terms. After World War II, large-scale projects (dams, tunnels, and so on) contributed to these quantitative approaches and to developing models to account for soil behavior. The methods in question rely on two components: at an elementary level, grains must be characterized; on a larger scale, it's a matter of measuring the texture and deformation of a mass of grains. Sixty years ago, a civil engineer, Pierre Dantu, was the first to observe the heterogeneous way forces are transmitted by means of *photoelasticity*, which we will examine in chapter 7. On the basis of these two factors, soil engineering could be progressively standardized and, in more recent times, automated.

Here, we won't go much further into the details a geologist could provide. Discussion will be limited to how physicists and engineers identify grains in terms of size, shape, surface state, mass, density, and interactions. These parameters correspond to the origin and history of the deformations that particles bear. Thus, a roundish specimen attests to journeys on wind and in water; successive collisions between one grain and others gradually efface irregularities. In deserts, winds carry the sand around for so long that dunes consist of practically nothing but rounded particles of uniform size. But on relatively large grains—starting with the fractures on pebbles—deformations can be traced back to the effects of tectonic or seismic movements.

Grains also belong to the world of vegetation. During the Devonian Period, for reasons that remain poorly understood, the first seed-bearing plants developed. So many plant groups have been discovered that their rapid development (which then continued during the Carboniferous Period) has come to be known as the "Devonian Explosion." The common ancestor of plants is algae; vegetation had great difficulty doing without water and needed to reproduce in aqueous media, often in a very short period of time. Under these conditions, the first true seeds appeared, containing an embryo, or germ, that could hibernate (that is, enter a state of dormancy) when conditions proved unfavorable. Seeds' ability to undergo intensive dehydration, which makes them resistant to predation and attacks by microorganisms and fungi, represents a key asset. Flowering plants, which appeared during the Cretaceous Period about 130 million years ago, made seeds that could remain inactive for quite some time. Lupin seeds, for instance, have sprouted after spending 10,000 years in the frozen soil of northern Canada. Nature is quite an inventor! Like a ship carried to and fro by currents and waves, the embryo is protected by an outer casing that resists shocks, predators, and changes in humidity and temperature. Some seeds have even developed strategies for covering more ground—for instance, the tufts of dandelions or maple samaras (which children often call "helicopters"). This natural use of hydrodynamics has inspired many researchers and designers. Still now, nature continues to show the way to those who wish to benefit from millions of years of improvement and optimization.

COUNTING GRAINS

In *The Adventure of Numbers*, the mathematician Gilles Godefroy tells the tale of Scottish warriors who, when setting out for battle, had the custom of placing a pebble on a pile. Those who made it back alive each removed a stone. By counting the remaining pebbles, they knew how many of their companions had fallen on the field. After all, *calculus* means "little stone"! To determine the approximate number of grains in a pile, one needs only weigh the whole and figure out the average weight of a single grain (assuming that the grains are reasonably close in size). The result can yield mind-boggling quantities. In *The Sand Reckoner*, the ancient Greek mathematician Archimedes set out to determine how many grains of sand it would take to fill the universe. According to the physicist David Louapre, there are nearly as many grains of sand on the Earth as there are stars in the universe; the numbers are big enough to make your head spin! On a more modest scale, one can content oneself with counting particles by passing them, one by one, in front of a beam of light, which they temporarily obscure—like people walking by a ticket window. It should be noted, however, that this method only works if the concentration of particles is low enough for our optical "doorman" to count them out individually. There are also other methods—optical ones, as a rule—for obtaining fuller information about an ensemble of grains. In particular, the way that light passes through a medium containing grains in diluted suspension tells us about their average density and size. When the grains reach one-tenth of the light's wavelength, the light is diffused, again, as in the case of milk or of a cloud. This is called *Rayleigh scattering*; it derives from the fact that the density and, in consequence, the index of refraction of a colloidal liquid fluctuates in space over time. Shining a beam of white light through a glass of dilute milk powder in water will reveal diffusion that produces a reddish glow on the optical axis and a bluish one transversally, because the different wavelengths of the source are sensitive to the scale of these fluctuations. This experiment simulates the diffuse radiation of the sun's light in the atmosphere. With particles of a bigger size, it's important to bear in mind that they cast a shadow when light is transmitted. This is called *Mie scattering*. Here, the wavelength does not play a major role, and the scattering does not affect color. A prime example is the effect of little particles of water in white clouds overhead.

MEASURING GRAINS

In order to understand the organization of granular matter, quantitative evaluation of certain constituent elements is necessary. We need to establish the grain's identity card. Although characterizing a grain in geometrical terms can sometimes require elaborate instruments, the principle of the process itself isn't complicated. In general, one proceeds by taking a sample of the population; as when dealing with a human being, weight, height, and girth are informative. Does the grain have a smooth surface? Is it bulky? Is its profile jagged? Whereas a round grain may often be characterized in one dimension, two or three parameters are necessary for grains that have the shape of a needle or a chip; on occasion, more parameters will be necessary if the grain is irregular. Since a complete inventory of factors would take up too much space, we'll discuss two basic principles to observe when measuring.

THE PERSPECTIVES OF PHYSICS AND ENGINEERING

For physicists, obtaining data efficiently is a cardinal principle. Introducing new parameters always means enlisting new instruments, new diagnostic tools, for recording information and calculating averages.

The second principle, which is closely related to the first, is managing data effectively. In an ideal case, a reduced and well-defined amount of information will closely approach the real facts by omitting characteristics that qualify as secondary. What is relevant for one problem might not bear on another one. For instance, whether it's polished or rough, bright or dull, any ball will fall at the same speed in a viscous liquid; that said, optical observation of the same process requires different tools. Engineers need to take into consideration all the details at work in a system to reach a precise quantitative assessment. They need to know grains and the assemblages they form as fully as possible, drawing up catalogs and general rules of their properties. Such charts are easy to manage using databases, and engineering now incorporates many parameters calculated by computer. On one occasion, when exploring the mechanics of threads of fiberglass insulation, we encountered an engineering program that incorporated fourteen parameters—which still didn't account for all the material's mechanical properties! That said, mathematicians can trace a line describing an elephant's profile by setting just five parameters; sometimes, a model can try to incorporate too many factors.

This, in schematic terms, is the difference between the approaches taken by physicists and engineers. Physicists look for hidden order by expressing the most general physical laws they can. Symmetry and scale can tell us a lot about grains' properties. In this book, we'll be trying to combine the insights afforded by both physics and engineering.

Instead of inventorying the geometrical properties of grains in detail, each problem we discuss will consider only what is absolutely necessary. The simplest model we'll be using has a single parameter—what could be better than that? Called the *hard sphere model,** or, more colloquially, the "billiard ball model," it works with undeformable spheres that all have the same diameter. The only parameter that admits adjustment is the number of particles per *volume unit* (or *surface unit* in the case of balls on a flat billiard table). Because of its universality, this model is appealing, but it won't do for many problems: try going down a slope made up of steel balls instead of rocks!

With this spherical grain as a reference point, it's possible to describe any other grain's deviation from regularity in terms of *elongation,** expressed as the ratio between its largest and smallest dimensions. Devices (granu-lometers) that capture simultaneous images of falling grains from differ-ent angles enable us to gauge this aspect of shape. Deviation may also be calculated by describing the smallest sphere enclosing the whole grain as well as the largest sphere completely included in the grain. For an almost-spherical grain, the ratio between the radii will differ little from unity.

GRAIN SURFACE

The smoothness or, conversely, roughness of a grain and its angularity are key parameters for understanding how it will interact with neighboring particles in a pile, and what their relative motion will be. Anyone who has ever been in a large, dense crowd and tried to move in the midst of all those people will easily imagine the jostling, resistance, and shocks at play between grains.

Surface condition bears witness to the voyage that grains or stones have made. In one particular case, this fact interests historians directly: prehistoric objects used as tools tell a story to anyone who knows how to listen. The uninitiated will confuse a sharpened flint (a rock made up of disordered aggregates of quartz and siliceous cement of chalcedony) and a stone that has been broken accidentally. However, the experts can reconstruct its preparation and use by reproducing the gestures performed by an ancestor long ago with tools (likely a wooden or bone hammer, or a rock); the process involves looking for the waste left in the course of fabrication, the surfaces of fracture, touch-ups, and alterations, as well as imperfections left behind after actual, practical use of them. In this way, a piece of flint can inform us about our forebears, who from the Paleo-lithic Period up to the Middles Ages used stones as weapons, tools, and ornaments (figure 2.1). Today, it's even possible to determine an object's age by studying how light elements such as fluorine, which initially were contained within the rock, have spread to the surface over time.

Examining the surface, then, is essential for understanding the nature of grains. An exemplary case is the uniformly rough surface that some-times results when a material is fractured. Whereas splitting a crystal yields flat planes on an atomic level, breaking a heterogeneous material generally yields rough surfaces at varied scales, on which the properties of surrounding areas depend; in this context, as we will discuss further on, the term *fractals** offers a key point of reference. Instruments have been developed for studying surface conditions and measuring roughness. They incorporate sensors directed at the surface of the material, and their vertical movement provides a record of it. The most remarkable device based on this principle is the *atomic force microscope*, an early version of which (the scanning tunneling microscope) earned the Nobel Prize for Gerd Binnig and Heinrich Rohrer, two Swiss physicists working for IBM; *atomic force microscopy** (AFM) allows one to detect, and even move, atoms and molecules that, magnified, look like a layer of billiard balls (see figure 3.2).

It is also possible to characterize mean surface condition indirectly. A more significant quantity of atoms can be adsorbed on an uneven surface than one that is smooth. At an extreme end of the spectrum, a surface with crevices on all scales displays highly irregular geometry of the fractal type. Such a surface can adsorb a monoatomic layer of gas. When the surface is rough, the smaller the grains are, the more they will manage to work their way into small cracks. (Thus, a fishing boat can find more mooring space available than an ocean liner!) Take, for example, a pile of powder consisting of spherical grains one micron in diameter. If one evaluates the total surface area geometrically, the result is 1,000 square meters for one liter of grains. The ratio is a thousand times greater for the same volume of grains of active carbon, or the zeolite powder used in industrial adsorbents for catalytic converters. This is due to the fact that these grains possess an extremely rough surface with many internal voids. The roughness of a grain governs its accessibility: increasing surface at a constant volume increases efficiency; the filters in kitchen exhaust hoods operate on this principle. In batteries, finely divided electrodes make it possible to increase the contact area with the electrolyte and, as such, improve overall efficiency.

Another technique for measuring roughness involves illuminating the surface with light at the wavelength λ: the light is split in all directions, that is, diffused, by surface irregularities. The distribution of luminous intensity as a function of the scattering angle provides quantitative information on the surface condition. Today, intense light can be produced by large instruments called *synchrotron** light sources (for instance, the

National Synchrotron Lab at Brookhaven, or the European Synchotron Radiation Facility in Grenoble), which emit eletromagnetic radiation ranging from the short wavelengths of X-rays to wavelengths in the visible spectrum. Modifying a wavelength is a bit like adjusting a gauge to determine the size of imperfections on the surface under examination. The latter may display smooth characteristics when illuminated by light at a broad wavelength, and rough ones when exposed to rays at the small values of λ.

In the following, such analysis will enable us to characterize contacts between grains, which may be described metaphorically as a meeting between two individuals: permanent or temporary; rough or smooth, hard or soft. Often, a third factor comes into play (water, a lubricant, or a glue).

Chapter 2 continues to explore the nature of grains by considering how they are made. That said, describing single grains and the contacts between them is not enough to understand the properties displayed by numerous grains in immediate contact with several others. We must be wary of a "reductionist" approach: knowing what happens on a small scale is not enough to account for what occurs on a larger scale. A packing of granular matter is much more than the sum of its parts!

2

GRAIN PRODUCTION

Around and around and around went the big sail
Turning the shaft and the great wooden wheels
Creaking and groaning, the millstones kept turning
Grinding to flour the good corn from the fields.
Alan Bell, "Windmills"

How are grains made? Often, grains of the right size aren't available for a project. Sand or gravel dredged up from rivers or the sea represent exceptions: they have many direct uses that don't require adjustments. A reasonable estimate of the quantity of this material used for making beaches (or just sandboxes), as well as for manufacturing in the glass and chemical industries, lies at more than fifteen billion tons. Especially now, it's important to avoid excessive exploitation of this resource to prevent coastal erosion.

Obviously, to obtain grains of the desired size, two opposing approaches are available: either start out with large objects and break them up, or use small objects and combine them. As we saw in our brief geological excursion, both of these processes occur in nature. Even though they're opposites, they may be described in parallel.

MAKING THE SMALL FROM THE LARGE: ANCIENT HISTORY

Let's go back to the Neolithic: human beings, trying to escape the ups and downs of climate and food supply, adopt a sedentary mode of life. All manner of crafts and trades appear, and technological advances multiply. New practices include ceramics, metallurgy, grinding and polishing stones, and farming. These activities all involve granular materials that either exist in nature or must be made by artificial means.

Neolithic, which means "New Stone" Age, is defined by the rise of agriculture, which requires tools for clearing forests and harvesting crops. Making implements from stones—a forerunner to milling—involved shaping rocks and rendering them more durable (figure 2.1). Results included axes and adzes with regular edges made from flint, basalt, or jade. These tools were ground over an extended period on stationary or mobile arrangements of sandstone or granite; in the process, quartz sand (which is extremely tough) and water were added at regular intervals; gradually, definition emerged until the surface became shiny—almost like a mirror. The Neolithic also witnessed the first mining efforts on a

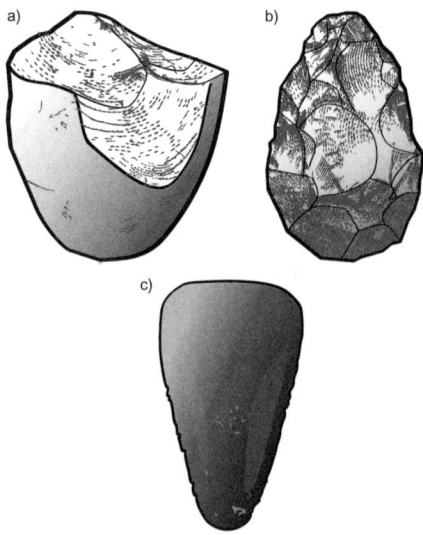

Figure 2.1
Prehistoric tools: a) chopper; b) hand axe; and c) polished stone axe.

grand scale. Still today, one finds remnants in deep wells, or even on the surface of the Earth, attesting to the existence of prehistoric workshops.

Farming—cultivating grains such as barley, wheat, and millet in Europe, rice in Asia, and maize in Mexico—fueled the rise of vast civilizations by ensuring regular and plentiful alimentation for growing populations. Agriculture also stimulated the development of new technologies for transforming grains into flour and winnowing out elements unsuited for consumption. Grinding tools could be transported—for instance, quern-stones: flat rocks with a center made slightly concave through wear, which stood on the ground; a second, oblong or cylindrical stone that fit in the hand (or was moved in the manner of a rolling pin) crushed and ground the grains. From this moment on, techniques and technology will keep improving.

*Ceramics,** or terracotta ("baked earth"), is a remarkable material that does not exist in nature. Our ancestors used it to make vessels, tiles, and pipes that would be impermeable and withstand fire. To this end, they had to find the purest *clay** and mix it with water so it could be shaped. By adding the right kinds of grains, they made an internal frame that would keep the clay from deforming afterward. Neolithic pottery contains sand and chamotte (fragments of used ceramics), as well as assorted powders such as ground-up calcite, shells, and pieces of bone. We will describe clays (chapter 9) and ceramics (chapter 10) later on, when our attention turns to the *consolidation** mechanisms of granular materials.

Like an alchemist or sorceror, the artisan extracted copper ore. With an array of implements including mallets, hammer-stones, indented rocks, grooving hammers, and mortars, bit by bit he crushed, sorted, cleaned, and, finally, enriched material drawn from the source rock. Ever since, these assorted operations have represented the first—inglorious, but indispensable—phase of metallurgy. In a furnace attached to bellows by means of a ceramic nozzle, our ancient forebear stoked the fire to reach a temperature above the 1,100 degrees (Celsius) necessary to melt copper. The process yielded an ingot. By adding some tin, he obtained bronze, fit for fashioning knives, needles, axes, and various kinds of jewelry. In one form or another, these same operations are conducted all over the world to the present day. With a little more sophistication, they're still used by modern craftsmen and in industry.

Figure 2.2 depicts activities related to mining. The illustration is taken from a classic, mid-sixteenth-century work on the subject: *De Re Metallica* by Georgius Agricola, "the father of metallurgy." Let's take a closer look at these operations.

We'll start with large grains, which can be broken by using a hammer and a hard surface. The industrial equivalent is to place grains in a ball mill, a time-tested device in cement plants. This equipment consists of a very long, almost horizontal rotating cylinder; as the large, metal balls knock about, the grains are crushed. Let's say we're interested in how a population of small grains—for instance, grains a hundred times smaller than the big blocks we start with—varies. This intermediate population will evolve by way of two opposing effects: on the one hand, when a large grain breaks apart, it will yield pieces smaller in size; on the other hand, a grain that starts out at a medium size will break into fragments too tiny for counting. The law at work—sometimes called a "golden rule"— allows us to characterize grinding operations, albeit without telling us anything about the mechanical and physical processes that bring about fragmentation.

Consider what happens in a coffee grinder. If, starting with a given mass of large grains, the grinding continues for long enough, the population of medium-size grains will increase at first, and then diminish— when most of the material has been pulverized (figure 2.3). The physical process of grinding determines just how fine the result will ultimately be. Moreover, a tiny grain size isn't always desirable. To understand grinding and sorting in large-scale processing, it's instructive to pay a visit to a mine or cement plant. As in other realms of science and technology, changing from the laboratory to large-scale industry—where the objective is to obtain "medium grains"—means reconsidering notions derived from model experiments conducted on a smaller scale.

The efficiency of operations in relation to the energy expended is crucial—a decisive factor in overall costs, even when raw material is inexpensive. In fact, the total energy cost for processing exceeds that of the whole transportation sector. The quantity of material to be broken down and ground up for the industrial operations involved in civil engineering is estimated to amount to one ton per year for every resident of industrialized countries.

Figure 2.2
Breaking rock for industrial use. Engraving from *De Re Metallica*.

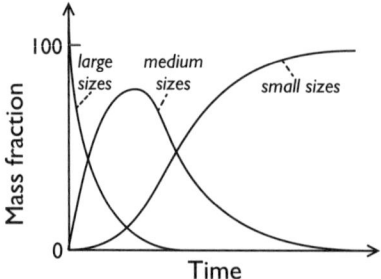

Figure 2.3
Distribution, as a function of time, of three grain populations as they are ground. The lower limit of grain size depends on the mechanism employed (like the settings on old-fashioned coffee grinders!).

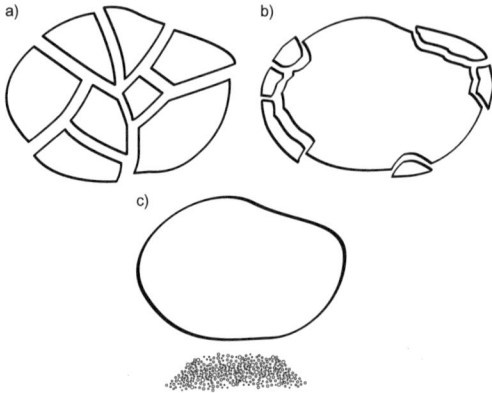

Figure 2.4
Grinding by a) fracture; b) flaking; or c) abrasion leads to different distributions of grains, in keeping with the material and tool employed.

FRAGMENTATION

Processes of *fragmentation* (figure 2.4) reduce an object's size, depending on its initial form, the hardness of the material, the defects it contains at the outset, and, of course, the intended use of the end product. Consider everyday operations in the kitchen and related settings. A cheese grater produces grains of regular size by way of attrition, or flaking, leading to the progressive diminution of the initial object's size. In the case of especially fine grating—say, parmesan—a powder results, and it's more appropriate to speak of abrasion. Butchers make cuts of meat by forcefully applying a sharp blade. (This is also how crystals are cut to have clean, flat sides—inasmuch as ambient pollution doesn't cover surfaces with a nanoscopic layer of debris.) Walnuts break under the effect of compression; separation that occurs naturally proves advantageous inasmuch as the shell opens into equal halves.

Energy efficiency is low in grinding operations. Cutting a crystal produces two flat surfaces facing each other, each one comprising atoms in periodic arrangement. Cutting a crystal in the desired directions is a bit like cutting a fabric along the lines of its weft or warp. The minimum basic energy required to break a chemical bond is some 10^{-19} joules—or one electron volt (the standard energy unit for chemical bonding)—between two atoms on either side of the cutting plane. If this value is multiplied by the number of atoms per unit area of surface—say, 10^{14} per square centimeter—this yields an energy level at 10^{-5} joules per square centimer. Relative to the given volume of material to be ground, this value is weak. In fact, the total expenditure, after creating a large surface by dividing the material over and over, will be significantly higher: at least ten times larger than the minimum energy at the outset. The finer the results of grinding operations, the greater the surface produced by a given initial volume will be—and the higher the cost.

The supplementary energy has become heat, a degraded form of energy. A small part has also transformed into acoustic energy: don't forget the deafening racket some grinders make! Given the total cost of energy in the grinding process, it's clear that all efforts to improve efficiency—even minimal ones that are far from approaching an optimal level—prove quite valuable for industry.

The mechanical study of how material breaks down involves an array of fracture modes, illustrated in figure 2.5. The starting point is an initial crack made in a given material. Traditionally classified as I, II, and III, these modes stand for how the fracture spreads between grains. They differ in terms of the orientation of forces applied to the extremity of cracks.

THE EXAMPLE OF MILLING
The history of milling illustrates advances in both technology and the use of new sources of energy. Mounted on boats, "floating mills" (or "ship

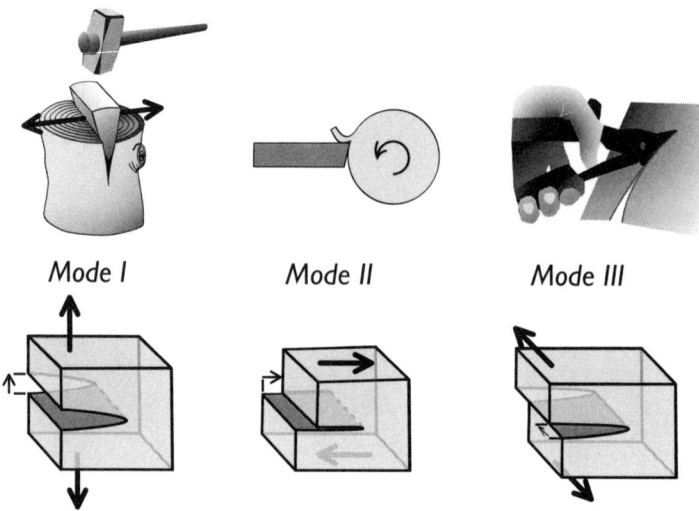

Figure 2.5

The three modes of fracture (opening, in-plane shear, out-plane shear) correspond to splitting, tearing, and shearing (and related operations). They are characterized by the direction of the crack tip in relation to force applied: traction for mode I, and shearing for modes II and III.

mills") used to be quite common; in Europe, they fell into disuse and disappeared only in the course of the nineteenth century. Their grindstone turned with the flow of water, exploiting the natural motor force of rivers (figure 2.6).

Grinding technologies evolved over time, but millstones were not replaced by metallic cylinders (which improve efficiency significantly) until the 1800s. Milling soft wheat, for instance, is a complex operation because a crease in the seed makes direct hulling impossible (unlike rice). Several steps of grinding and sifting are required to open the grain and scrape out its contents by removing the endosperm from the bran and the germ. To avoid confusion, a milling process diagram is used—a veritable algorithm describing the sequence of steps. Hereby, distinctions are made between hulling, when the grains pass through a series of "roller mills" (corrugated cylinders made from chilled steel) to be crushed into pieces, and reduction, which uses smooth cylinders to take advantage of different levels of *elasticity** between the kernel and residual husks in order to separate them. The final stage involves passing the wheat through a

Figure 2.6
Grindstone made from sandstone, found at the Gallo-Roman excavation site in Arnouville.

series of smooth cylinders with an increasingly small distance between them; this produces a finer and finer product. At each step of the way, plansifters—tiered sets of sieves—separate particles by grain size and steer particles of the wrong dimensions back to the appropriate grinding tool.

THE IMPORTANCE OF GRANULAR MEDIA IN PROCESS ENGINEERING

All these operations for reducing grain size are still in use. The technology has improved, but no real revolution has occurred. Although they may seem crude and fit for galley slaves, such operations represent the most important means for process engineers to obtain as much surface area as possible, in order to make as much solid matter as possible react with fluids. Indeed, the subject is so important that Antoine Lavoisier's foundational study of chemistry (figure 2.7)—published in 1789, the year the French revolution ignited—devoted an entire section to it.

NATURAL PROCESSES OF FRAGMENTATION AND EROSION

The motion of materials entails the degradation and production of grains. The greater part of the grains we use comes from processes in nature. However, the study of granular materials also holds major interest for astronomers, especially planetologists. The Moon's surface—in

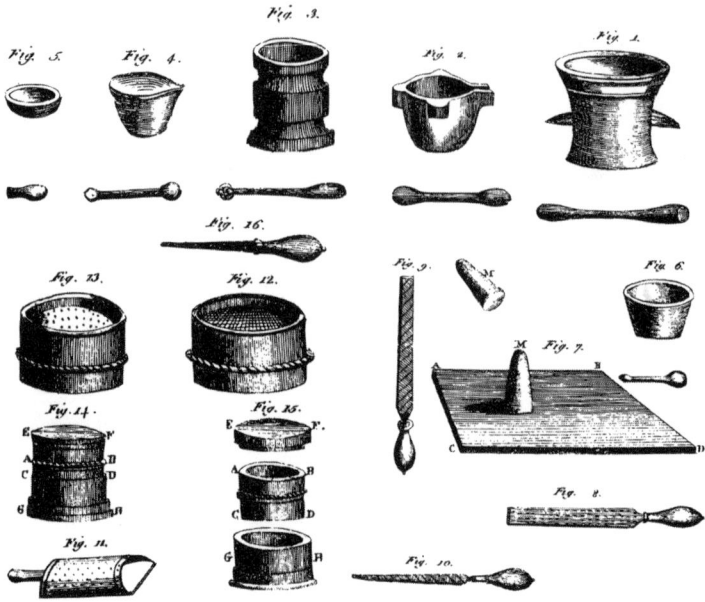

Figure 2.7
Extract from Antoine Lavoisier's *Elementary Treatise of Chemistry*, stressing the importance of dividing matter.

particular, the tiny grains found there—command attention. How can we
account for the presence of so many small pieces of rock and dust on an
inert body, where conditions seem calm—without volcanos or rain? The
answer comes from space: the lunar soil is constantly being bombarded
by meteorites. On impact, larges grains break up into smaller ones. Sometimes, having already reached an extremely small size, they are stuck back
together and form aggregates as a result of shock. The average size of such
grains of dust is quite small, on the order of fifty microns. This circum-
stance presented major difficulties for the Apollo missions. At this scale,
without moisture, and with gravity six times weaker than the Earth's,
electrostatic forces are strong enough to make the grains stick to space-
craft and work their way in everywhere.

On Earth, weathering and wear prove to be much more complex, given
the many forces at play. Boulders dislodged from the sides of mountains
are broken, worn down in the heart of glaciers, and carried by rivers and
winds. In cold and wet climates, water that works its way into the rocks

acts as a wedge when it freezes, causing fissures to spread. In a comparable manner, temperature variations of more than 40°C between day and night—in the Sahara, for instance—cause thermal expansion, which also creates cracks. When minerals and salts crystallize, they exert a *pressure** that can also weaken and fracture rocks.

Rocks and stones are rarely homogeneous. Granite, for instance, is composed of pieces of quartz, feldspar, and mica that have crystallized into small grains deep in the Earth. Though highly resistant, this conglomerate degrades over time. When they come into contact with air and water, mica and feldspar transform into other minerals—clay, for instance—or extremely fine grains that can be less than a micrometer in size. Now, the granite is just brittle rock that breaks apart and winds up in myriad pieces at the foot of the source rock. This acidic soil, which acts as a filter, is formed from crumbly, granitic sand called "gore," which is often colorful and rich in trace elements—and highly esteemed by Beaujolais winemakers.

Chemical processes can also be more aggressive and dissolve rock and stone directly. The karst caves found in the Causse limestone plateaus to the south of the Massif Central in France (which contain stalactites and stalagmites) represent a spectacular manifestation of such degradation. On the surface, the gradual dissolution of the stone combines with the pressure of water in joints, faults, and veins to make a landscape in tormented, ruin-like relief.

Many other chemical reactions can occur, for instance, hydrolysis, oxidation, hydration, and dehydration. These reactions, which are often complex, can encourage minerals to dissolve; they can also tranform one chemical species into another. Vegetation likewise plays a role in changing rock and soil; tree roots work their way into boulders and break them apart; by capturing ions in solution, they cause a chemical disequilibrium that entails further reactions. Finally, bacteria contribute to the decomposition of minerals, sometimes even making them precipitate into solutes that cement little particles of sand to form aggregates the size of large pebbles. Weathered and fragmented stones break free from the source rock. In smaller and larger sizes, they travel down rivers, as far as the depths of the sea. Such erosion heightens the effects of *shearing*,* and it can take different forms. In arid regions, where vegetation no longer protects the soil, the wind blows grains into dunes. Water run-off carries large rocks that, in the course of their journey, turn into smaller and smaller grains conveyed by rivers. In this way, the Rhone—just on its own—carries almost ten million tons of granular matter to its mouth every year, the result of rocks breaking up and erosion from one end to the other.

MAKING THE LARGE FROM THE SMALL: GENERAL REMARKS ON AGGREGATION

There are myriad ways small grains attach to each other to make bigger ones. Depending on how they are piled, the resulting grains will be more or less compact and form larger or smaller objects. The following explores only the aggregation of tiny numbers of grains, which, together, make a single macrograin. The particles obtained in the first two groups have a maximum size on the order of a micron.

Aggregation represents an operation symmetrical to fragmentation. If the flow of time were reversed in the diagram on figure 2.3, it would describe the evolution of the population cluster from small to large. The Polish physicist Marian Smoluchowski first formulated this law governing aggregation, which he applied to colloids whose *Brownian motion** he was investigating.

FROM CRYSTALLINE AGGREGATES ...

Crystals represent the result of ordered stacking—which sometimes proceeds quite slowly—of atoms that form periodic networks (which we will discuss in chapter 3). The size of crystals that can be synthesized by a periodic stacking of atoms may vary between crystalline microedifices visible only through a microscope to the gigantic, natural crystals in the caverns of the Naica Mine discovered in Mexico in the 2000s, where crystals of selenite gypsum reach a meter in diameter and more than eleven meters in length. On an intermediate scale, piles of spheres or grains constitute *metamaterials** (*meta* means "beyond" in Greek), which are now beginning to find practical applications and have properties distinct from those of ordinary solids made of piles of atoms.

... TO GRAINS OF COUSCOUS

Here, too, normal activities shed light on the mechanisms at work. Have you seen semolina being rolled into grains of couscous? The traditional way to make it in the Maghreb requires manual skills it takes people a few years to master. After soaking one's hands in salted water, one places a few drops of water in the middle of the plate, without directly wetting the semolina (which would make it lumpy). By means of a quick turn of the hand—to prevent coarse aggregates from forming—the semolina is then moistened; a continuous, rolling motion between the palm and the plate ensures that the grains of couscous will grow larger and larger by gathering together with small particles of dry semolina (figure 2.8). The next stage involves sifting between two sieves, one placed on top of the other. The lower sieve, which has a finer mesh, catches calibrated grains of couscous, allowing particles that are too small to fall through (to be recycled). The sieve on top keeps large pellets, which eventually find another use, such as baked in a cake. Finally, the couscous is dried in a ventilated room or a basket in the sun.

Figure 2.8
Photograph taken by scanning electron microscope: agglomerates of couscous at an early stage of growth.

COLLOIDS

Suspensions of stable particles of gold made up of metallic grains only a few nanometers large were known to medieval alchemists; they were used to color stained-glass windows in cathedrals and for illuminated manuscripts. In the mid-nineteenth century, Michael Faraday conducted experiments with such suspensions and produced samples that have been preserved to this day, without any change! However, by carefully adding a product that neutralizes the charges holding grains apart and ensuring the stability of the colloidal suspension, small aggregates form like the ones presented in figure 2.10. These particles of gold dust have stuck together to make a disordered structure that is loose and ramified, with "fingers" of varying sizes. The treelike structure of this colloidal grain is easy to understand: particles of gold in solution add themselves to it by way of a winding *random (or Brownian) walk,** which results from their thermal agitation. It's relatively easy for them to attach to the "capes" of the rugged "island." By the same token, it would be difficult for them to reach the bottom of a "bay" without getting stuck to a "bank." This process amounts to a veritable point effect—as if the colloidal grain were a conductive material in a vacuum drawn toward an electric potential: the electric field is strongest at the tips of the grain assembly.

FRACTAL FORMS

An essential characteristic of these colloidal objects is that enlarging (zooming in on) one part reveals the same average structure that the aggregate displays on the whole. This is the defining feature of what is known as a *fractal.* In fact, many objects have the same structure even when observed in highly different experimental situations. They can also be produced on a computer—for instance, by determining the arrival times of a series of grains on the basis of the interplay of colors. The results of such simulations are quite elegant in aesthetic terms, and often go beyond the scope of actual scientific experimentation.

This ensemble of grains of colloidal gold will serve to introduce fractal forms, which represent a key tool in the science of *disorder** (figure 2.9). But first, Jean Perrin has the floor. In the dazzling introduction to *Les Atomes* (1905), he wrote:

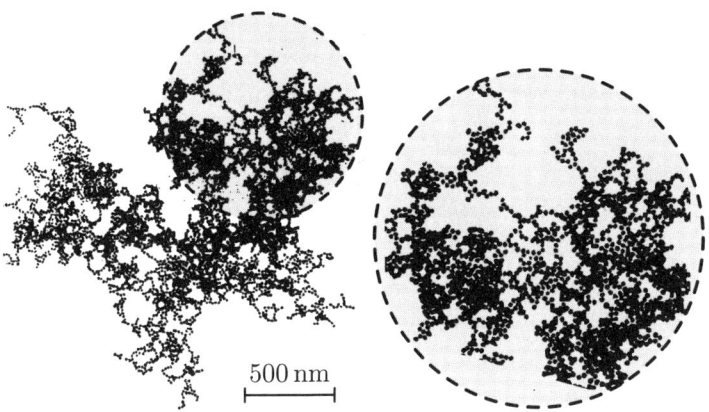

Figure 2.9
Electronic microscope 2D image of a colloidal cluster of gold, composed of individual particles a dozen nanometers in size. Its structure is self-similar at different scales, as indicated by the enlarged picture of the aggregate (on the right). Its fractal dimension D_F lies between that of a line (D = 1) and that of a surface (D = 2).

Consider, for instance, one of the white flakes—a colloid—that are obtained by salting a soap solution. At a distance its contour may appear sharply defined, but as soon as we draw nearer its sharpness disappears. The use of a magnifying glass or microscope leaves us just as uncertain, for every time we increase the magnification we find fresh irregularities appearing, and we never succeed in getting a sharp, smooth impression, such as that given, for example, by a steel ball.

Nature provides countless examples of the same, average structure recurring on different scales.

FROM DUST TO STARS

"The small is great, the great is small; all is in equilibrium in necessity." In the visionary capacity of the poet, Victor Hugo describes the universe as an arrangement to be understood as a unified whole, from the tiniest part up to the most immense structure.

The story of a given planet starts out in proximity to a nascent star, surrounded by a cloud of gas and dust. Just as a figure skater will draw in his or her arms to transform into a spinning top, rotation accelerates as the cloud collapses on itself under the effect of gravitation. This phenomenon leads to the emergence of a protoplanetary disc comprising various simple elements; it also includes metallic and mineral compounds that accrete in small grains, in keeping with the mounting temperature closer and closer to the star. Near the disc's interior, where it's hot, refractory objects are found: rocks. Where it's colder, more volatile elements can condense and become solid—for instance, water in the form of ice. In our own solar system, the so-called "frost line" (alternatively, "snow line" or "ice line") is found a little before Jupiter; up to this threshold, simple molecules will condense. This line marks the limit between small telluric planets close to the Sun, which have managed to retain only a limited quantity of lightweight molecules (hydrogen and helium, for example) and nebulous giants that have captured them in the form of liquid or ice.

NASA's aptly named *Stardust Mission* collected samples of interstellar dust floating in the void. It seems clear that these minuscule fractal aggregates a few hundred microns large are the original material from which our planets were born. At present, the world's largest computers are being used to test out hypotheses concerning the physical mechanisms that led to their accretion. Scientists think that tiny specks of dust collided and progressively increased in size until they formed rocks. These small bodies, which resemble asteroids, are called planetesimals (figure 2.10). Once

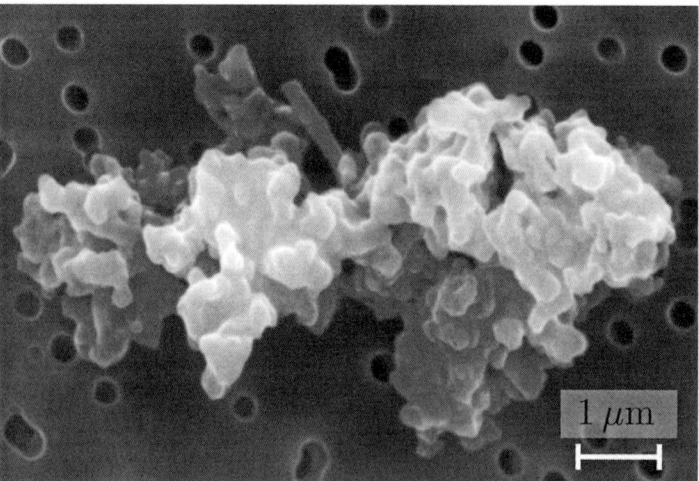

Figure 2.10
Cosmic dust a few microns in size. Herbert A. Zook, from NASA, has estimated that forty thousand tons of cosmic dust fall to Earth every year.

they're big enough to attract material by gravitation, they are capable of capturing dust that doesn't lie in their direct trajectory. In our solar system, there once were probably several hundred little planets, or planetary embryos.

SORTING OPERATIONS

"Separate the wheat from the tare." The words of the apostle Matthew refer to working the soil: the need to remove weeds from fields of wheat so their toxic grains won't spoil the harvest. Still today, producing seeds and fruits requires many stages of separation. Sorting operations employ an array of methods, and often several at once. First, a combine harvester is used—the emblem of the rapid evolution of agriculture, and one of the greatest advances in the mechanization of farming. Before being conditioned, grains separated from ears of wheat by threshing must be minutely sorted.

Mixing compounds for a high-performance *concrete** or porous asphalt as well as manufacturing a compound of so-called *Galenic pharmacy* so it will dissolve at the proper rate are other examples of operations that call for carefully monitoring the size of grains (or range of sizes). The finished product depends on the judicious choice of grain proportions just as much as their chemical composition. We will now examine, in general terms, a certain number of sorting operations and classify them in keeping with the property that enables grains to be distinguished and separated. A discussion of filtration (which separates large grains from small ones) in relation to porous environments is reserved for the end of the book.

USING DIFFERENCES OF GRAIN GEOMETRY

Separating a grain of sand, a piece of metal, a small insect, or a blade of straw from a grain of wheat before turning it into flour can prove to be quite a complex operation. Seeds may vary greatly, and processing them often requires specific technologies. Canola seed is round and smooth, whereas oat seed is long and pointy, and cat grass has long, stubbly spikes. An imposing variation of shapes exists among the thousands of kinds of rice, wheat, and maize.

How, then, is the right grain sorted out? One can take advantage of differences between the physical properties of the desired grain and the impurities in the mixture. A magnetic field will draw out iron residues, for instance. An electric field proves useful to separate differently charged particles thanks to electrostatic force. Technologies of this kind may be applied on a large scale for recycling industrial waste; they're also good for promoting a corporate image of protecting the environment.

There are so many sorting devices, it would be impossible to list them all. Indented cylindrical separators, for example, classify grains in terms of shape (oval, long, or round grains). These devices consist of a rotating drum lined with little cavities whose shape is close to that of the grains to be selected. Like pieces of a puzzle, the grains will tend to stick inside these sockets. The movement of the siding then draws them along until they fall into a hopper, while undesired elements are left at the bottom of the drum.

The cleaner-separator—the modern descendant of the winnower, or fanning mill—relies on lift to separate lightweight refuse (bits of husks or straw) by means of a gust of air. In general, this device employs fixed sieves with circular or elongated perforations that make it possible to eliminate denser particles as a function of shape. Gravity separators work on the principle of density to remove unhealthy, parasitized, or broken grains; a layer of grains circulates slowly on a vibrating, porous surface at a slight angle, with air blowing from below: the least dense particles are held in suspension at the surface of the bed and gradually separate as grains move about on the table.

FLOAT OR SINK

In mineral processing, one of the most important industrial technologies is flotation, which separates ore from gangue by taking advantage of these phases' differing wettability. Foaming agents are employed that are adapted to one phase or the other and make it float, facilitating separation. The process allows a material containing 1% copper ore to be enriched up to about 30%; the part that is eliminated contains no copper at all. More physico-chemical agents are used for flotation than for better-known forms of surface treatment, such as washes and detergents.

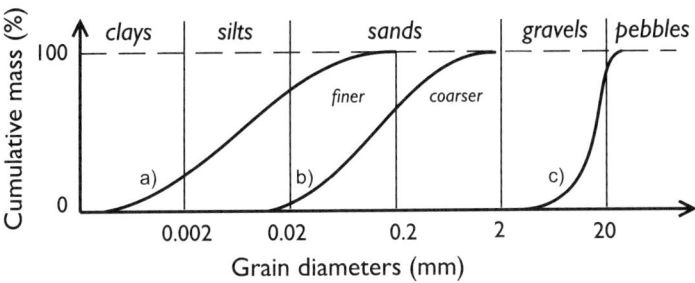

Figure 2.11
Grading curve of a) loamy-clayey soil; b) sand; and c) gravel.

A given granular material contains particles of varying size distribution (figure 2.11). A series of sieving operations makes it possible to separate a mixture of grains and organize it into "classes" (figure 2.12).

PONDERAL EFFECT

Pour a handful of garden soil into a transparent vessel filled with water then give it a shake to disperse the soil and leave it to rest. Obviously, the large grains will fall to the bottom of the vessel first. If the mixture is initially homogeneous, size distribution will vary over time as *sedimentation** proceeds. If, with a pipette, a series of samples is taken at a given depth—or, alternatively, at different depths at the same time—the same dynamic evaluation of sedimenting grains of different sizes will result.

When grains are smaller than, say, a tenth of a micron—as is the case for colloidal grains—the effect of gravity is insufficient to ensure sedimentation and to overcome thermal agitation, which tends to keep the concentration uniform; moreover, the water in the container will stay grayish because the clay particles remain in suspension due to Brownian motion. We can also place our sample in a centrifuge, which spins the tube around a vertical axis (up to several thousand revolutions per second in the case of an ultracentrifuge) to reach an apparent weight 10,000 times larger than the object normally possesses, which renders the effects of thermal agitation insignificant. Today, a high-speed centrifuge represents an indispensable component of every biochemical laboratory.

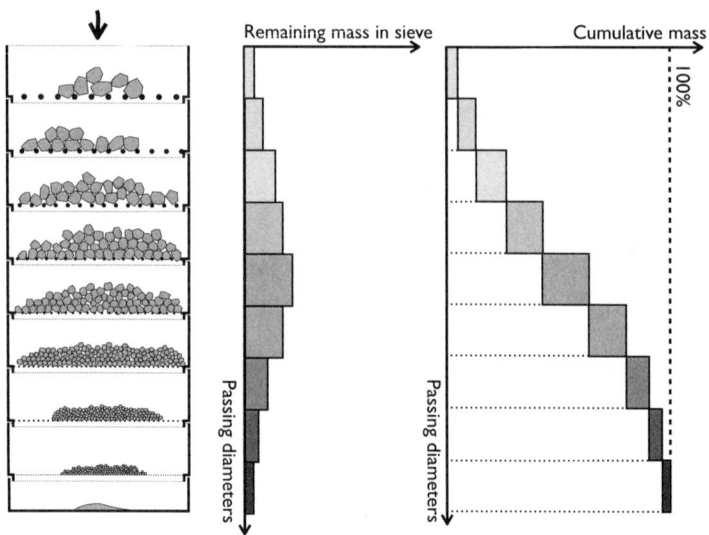

Figure 2.12
Separation by size through successive sieving. This method is also used to determine the grading curve, by weighing the mass each sieve retains.

DIFFERENT MOBILITIES

Biologists and chemists have a wide array of technologies for separating the larger Brownian molecules. These means also apply to small grains of matter, for which the effect of thermal agitation is appreciable; blotting paper, over which a suspension of particles spreads out, provides an example of molecular separation by *diffusion*, without external force. What is more, in response to an external force field such as an electric field on a charged grain, or gravitation due to their weight, these particles display *mobility,** which is the ratio of their speed to the applied force. (Note that in the case of heavy grains, it's acceleration and not velocity that is proportional to the force applied.) Since Einstein's work on Brownian motion in 1905, it has been known that spontaneous *diffusivity,** as well as mobility in response to force, are proportional. A combination of various fields—for example, gravitational effects, electrical fields, and flow regimes—is often used to separate grains. We won't inventory all these techniques, which generally concern grains smaller than those under discussion here.

ADAPTING THE SIZE DISTRIBUTION OF GRAINS

Many applications seek to retain only grains of a particular size. That said, there are many cases in industry when it is necessary to obtain a range of sizes. This operation is known as particle-size distribution. We will examine the benefits it affords later on. Here, we will provide an example that, while familiar, perhaps has not been examined closely enough.

The Example of Ballast   The ballast on railways seems to be just a pile of stones placed under the ties that support the rails. All the same, this non-pollutant material proves indispensable; to this day, no replacement has been found. Its principal function is to ensure the connection between the rails and the ground, without bending under the considerable weight of trains. It also plays a key drainage role, ensuring the track's stability in the event of heavy rain. Finally, it serves to soften vibrations that cause unwelcome noise, and it prevents vegetation from growing.

Using rocks for ballast represents the fruit of many years of experience and trial. Even though railways have been around for more than two hundred years, ballast is still the object of active research (see figure 4.5). Durable, calibrated pieces of granite, diorite, and sometimes siliceous limestone are used, depending on climate and availability. Operations at ballast pits follow extremely precise specifications concerning size, resistance to wear, and fractures resulting from weight or freezing. There are even special quarries for the ballast used for high-speed trains; the material they provide needs to be solid enough to withstand the high frequencies generated by trains racing at three hundred kilometers an hour.

Ballast may settle over time, and even wear down or become polluted by dirt and *microscopic** particles. Then it loses its initial grain size and no longer functions correctly. In such cases, a tamping machine is used, which, by means of vibrating metal stakes planted on either side of the ties, makes it possible to compress the ballast under the tracks and adjust the rails' geometry. Over longer intervals of time, it is also necessary to replace the used ballast with new material. The thickness of the ballast layer, the frequency, clamping force, and penetration depth of tampers, as well as the amplitude of vibration provide the essential parameters for this operation, which is typically performed at night to avoid interruptions of service. For a thousand meters of tracks, it takes 120 tons of rails, 1,600 ties, and 2,500 tons of ballast!

Concrete   Filler, sand, gravel, ballast ... these are all different names for describing the various sizes of grain that go into concrete and mortars (figure 2.11). An appropriate sand will contain fine, medium, and coarse grains. If only sand from dunes were used, the concrete would exhibit poor resistance; the dwellers of desert regions only use the surface layer—the crust—because it is much more heterogeneous. The significance of such gradation has been known for quite some time, but recent advances in so-called high-performance concrete have been achieved by adding extremely fine grains (silica powder, for instance) to the overall mixture of sizes; in this way, porosity is greatly reduced and mechanical properties are appreciably enhanced. We will discuss this point in detail in chapter 9.

We have completed our initial stroll among granular matter of different kinds. Going forward, we should bear in mind a few general observations and principles that inform the chapters to follow. Each grain of matter has its own properties and history. Like an elementary adventurer, it can travel through the air and water, or over land. We have seen how to recognize grains according to their properties, size, shape, and surface condition—but without providing an excessively detailed catalog. We have also considered the manufacture of grains starting with larger or smaller objects—two opposing procedures that run in parallel. Finally, we have discussed how to sort grains. As yet, we haven't lent much attention to the specific problems confronting an engineer, which justify assorted operations of production, description, and separation. The variety of areas of application is vast, and further characterizations of grains must be considered in light of their specific uses.

3

PILES OF GRAINS

Form and matter are connected, and the one preserves the other. Form, then, is an integral of time; it absorbs, as it were, preceding ages into the present.
Paul Valéry

In this chapter, you are invited on a stroll—or, to be more precise, two strolls—among piles of grain. Our construction game will make use of marbles, atoms, and, in between, colloidal grains as building blocks. We will meet both ordered piles (periodic assemblies of particles, as in crystals) and disordered piles. But our journey will also lead us to consider spatial dimensions: starting from a single grain (in zero dimension, as it were), materials organize in chains, layers, or volume (that is, in one, two, and three dimensions). Current applications extend from grains the size of an atom all the way to macroscopic scales; in between lie nanometric scales that new technologies of microfabrication have disclosed. Later on, we will address recent applications of these metamaterials. We should note that the properties of systems depend on the scale at which grains are found—and that their applications depend on them, as well.

THE ARCHETYPE: BILLIARD BALLS

Piles present an extreme range of diversity. At a market stall, rows of well-sorted oranges stand arranged to form a regular pyramid (figure 3.1a). This is not the case for the pile of potatoes lying next to it (figure 3.1b). A scree is also an irregular and slanted pile, made up of stones of diverse shapes and sizes. Cairns (figure 3.1c) commemorating an event, marking a grave, or showing a path are also piles of irregular grains, but they have been designed for an express purpose. A bag of billiard balls or marbles can be considered as a model system for such disordered piles.

Figure 3.1

Assembly of grains in volume: a) well-stacked oranges form the beginning of a crystal lattice–shaped structure; b) a pile of nonspherical objects distributed carelessly will generally prove to be disordered; and c) the stones composing a cairn at a Swiss mountaintop are well piled but bear long-range disorder.

The various states of organization of grain assemblies that we will learn to recognize in this chapter provide the basis for understanding the phenomena to be described a bit further on. Variations introduced by shaking, compression, or welding grains together entail a wide range of geometries and behaviors. Additional effects arise from differences of shape and size.

For ages, packings of spheres provided a model for physicists studying crystalline solids. Today, they are also used to characterize disordered and amorphous structures. This model can prove inadequate when dealing with practical applications. In due course, we will see how to enrich our understanding of granular materials in context. But all the same, billiard balls or marbles provide a good point of reference for what physicists call *hard spheres*.* Well before modern theories were developed, scientists used regular piles of uniform size to represent the microscopic structures underlying the angular shapes of crystals. These models were developed long before the reality of atoms was accepted in the twentieth century, even though the Greek philospher Democritus had anticipated the discovery some 2,500 years ago, and the Roman poet Lucretius had popularized the notion in *De rerum natura*. Today, high-resolution electronic microscopy and atomic force microscopy can show us the surface of a crystal displaying atoms distributed regularly, just like an array of spheres on a level surface (see figure 3.2).

Figure 3.2
Transmission electron microscope image of a hexagonal monolayer of graphene. The atoms look like spheres; the distance between them is 0.14 nanometers.

The model based on a smooth, rigid ball is appealing because the stacking structure this means can achieve doesn't require measuring any parameter: whether cannonballs, oranges, or ball bearings are employed, the same organization is obtained, independent of the shape of the vessel holding them, provided it's big enough relative to the size of individual components. Here's a simple example to satisfy your curiosity: fill one vessel with microscopic beads of glass (the same kind mixed with paint on roads, so the markings will be reflective) and another container of the same size with large beads: one can add the same amount of water, more or less, to fill the two vessels, despite the fact that the sets of beads are different in size.

What are the characteristics of ping-pong balls used to play table tennis? A ping-pong ball is hard. In contrast, a tennis ball is deformable—and, moreover, covered with felt. These two characteristics illustrate the different ways of bouncing and the effects of interaction that prevail on a small scale in atomic structures—and, on a somewhat larger scale, in colloids (which we already presented in chapter 2 and will consider here in crystalline form). On the microscopic level, the effect of elasticity is due to clouds of electrons that surround atoms and forge chemical bonds. In colloidal grains, the "felt" covering comes from chains of polymers that can be grafted on, like hairs on one's head. This covering is responsible, for instance, for the stability of the grains of lampblack suspended in India ink. One energy parameter that accounts for interatomic forces, as well as for our ball, comes from the elasticity governing deformability. These effects belong to the realm of physical chemistry and colloidal science (concerning tiny grains suspended in a liquid with physico-chemical interactions); they will not be discussed at any length here.

To stick to the model of a tennis ball, the state of wear can be described by a parameter without dimension: the ratio between the thickness of the felt, δ, to the ball's radius, R. As the ratio δ / R decreases with wear, the stacking potential of the balls changes; eventually, it reaches zero, when the tennis balls are as "bald" as ping pong balls. In fact, for a pile of tennis balls (or soft spheres), two parameters are required: a ratio between the size ratio δ / R and an energy parameter tied to the balls' deformability. By using a dimensionless ratio between two quantities of the same nature—the radius R of the tennis ball and the thickness δ of its felt—the principle of economy is preserved because two variables, δ and R, are reduced to just one number: δ/R. *Analogous* ratios (in fact, the term is a pleonasm—in Greek, *analogia* means "proportion") often underlie the reasoning of physicists, even though textbooks are not always so clear about this fact.

ORDERED PACKINGS

Discussion here will be limited to the archetypal case of a pile of hard spheres with the same diameter—that assortment of billiard balls without a reference scale for size or energy—that has inspired many natural scientists, physicists, mathematicians, poets, and philosophers. This aggregate of spheres fits well with the words of Valéry quoted at the beginning of the chapter. We will discuss the order of atoms within crystals and proceed, step by step, to the complete disorder exhibited by atoms in a glass. To begin with, our attention falls on 2D models.

SPHERES ON A PLANE

In addition to conical grain piles, it is necessary to consider situations where organization occurs on a plane, in two dimensions. Doing so allows us to understand the geometrical effects of order and disorder more fully. A monolayer of tiles in a game of checkers or coins of the same size collected on a slightly inclined plane will spontaneously form a regular, "periodic" structure (figure 3.1b) (which, as we will see, is not the case for piles in three dimensions). These piles can be described as periodic lattices, which one can picture as a piece of wallpaper reproducing the same patterns periodically. Here, the elementary motif consists of equilateral triangles of spheres. Starting from an initial cluster of three plates, the simplest form of two-dimensional periodic packing is obtained: a triangular stack (figure 3.3a).

It is also known as hexagonal packing, albeit incorrectly, because six discs forming a regular hexagon would be in contact with a central disc; in order to create a stack of this kind (a classic structure of kitchen tiling), the plate in the middle of the hexagons would have to be removed, as shown in figure 3.3b.

a)

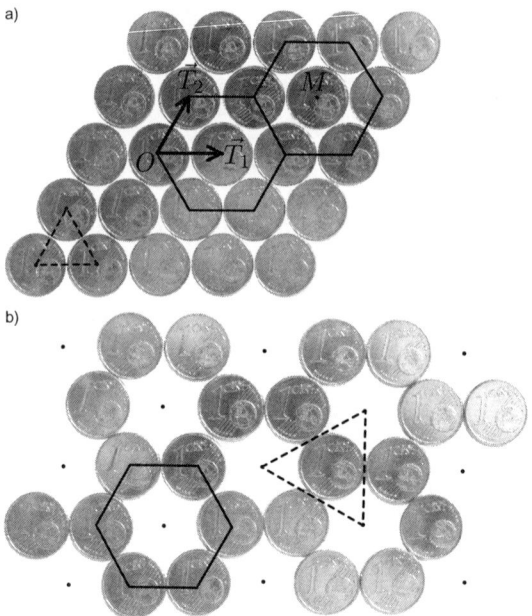

b)

Figure 3.3

a) Tessellation of discs forming a triangular lattice. With two basic vectors, $\vec{T}_1$ and $\vec{T}_2$, all the centers of discs (i.e., all the vertices of the lattice) can be obtained by a simple relation, $\overline{OM} = n_1\vec{T}_1 + n_2\vec{T}_2$, whereby n_1 and n_2 are integer numbers (in this example, $n_1 = 2$, and $n_2 = 1$). b) Tessellation of a hexagonal lattice. Centers of hexagons form a triangular lattice. Likewise, in case a), the centers of triangles form a hexagonal lattice. Consequently, the terms "triangular" and "hexagonal" are often confused.

Hexagonal Packing of Graphite Sheets

Hexagonal packing (or lattice) is the arrangement of carbon atoms gathered in the lamellar structure of graphite seen in figure 3.2. Each time a pencil makes a black line on a piece of paper, the mark is due to the deposition of several layers of material that consist of the parallel layers of graphite (like the pages of a book) that have been left behind. With a little luck, a piece of tape applied to a flat graphite surface will remove a single layer of carbon atoms: this is called *graphene*.* Two-dimensional physics thus is a reality! This simple experiment recently earned two physicists the Nobel Prize. The nanometric flakes can be rolled into a tube or enclosed in spherical surfaces. A whole field of application has exploded, with many everyday uses (such as carbon-fiber materials and sheets); countless more are still to come.

Let's get back to geometry. Tessellation and packing—the province of architects and geometers—have a long history. The astronomer Johannes Kepler has the distinction of having formulated the problem of crystallography in these terms. His considerations are found in an opuscule, "The Six-Cornered Snowflake: A New Year Gift," presented in 1611 to the court counseler of the emperor of Bohemia and Hungary. This text, which the author describes as "small and insignificant," answers the question, "Why do snowflakes fall with six equal corners?" Kepler's analysis likens this pattern to the network of cells in honeycomb and to a flat arrangement of identical spheres, evoking "the archetype of beauty found in a hexagon."

The lattice is like the skeleton of the pile; the spheres represent the flesh, as it were. The simplest distinguishing trait of this "flesh" is its compactness or, in more technical terms, its *packing fraction,** the portion of the surface that is occupied by discs in contact with each other. Basic calculation reveals that this packing fraction is $À/(2\sqrt{3})$, which is approximately 0.91 for a *triangular packing*. This represents the maximum packing fraction that can be obtained on a plane with discs of the same diameter. The surface area of voids between discs, the *porosity** ϕ of the structure, is thus 9%. A *square pile* is achieved by placing the spheres at the corners of a square whose sides are equal in length to the diameter, 2R, of spheres. Its packing fraction has the value of $C = \pi/4$, which is the ratio between the surface πR^2 (one of the discs) and the square measuring 2R per side, or $4R^2$ (figure 3.4a). This value, 0.79, is noticeably lower than that of the triangular arrangement. The number of contacts between the spheres and their neighbors, or *coordination number,** goes from four to six in the shift from a square lattice to a triangular lattice. Only rarely are regular square packings found in nature. Such structures are unstable as they tend to yield a more compact filling of space with an applied strain (figure 3.4b).

SPHERES IN VOLUME

So how do we get from this description of spheres on a plane to an ensemble of spheres in volume? Two approaches are possible. The first is to pile crystalline layers in two dimensions on top of each other; doing so yields

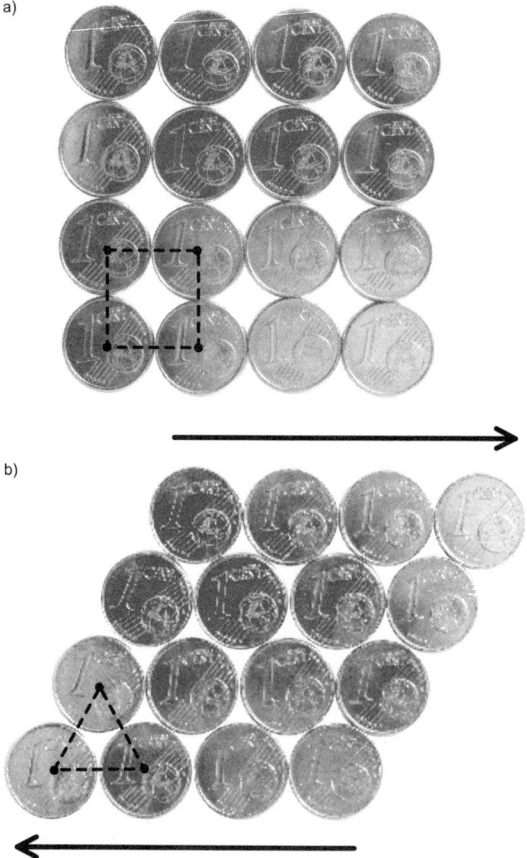

Figure 3.4
a) Square lattice and corresponding tessellation. This lattice is less compact than a triangular lattice, and not very stable. b) Sliding the layers of the square lattice in parallel to their direction yields more compact triangular stacking.

structures of crystals in volume. The second approach, which we will adopt later on, starts out with the simplest possible local structure: a pyramid of four spheres in contact, whose centers form a tetrahedron; this leads to disordered structures, which are those most commonly encountered in nature. We will now explore these two approaches in detail.

Let's start by studying ordered structures in space (figures 3.5 and 3.6), starting with a plane of spheres, *A*, forming a regular, triangular lattice. The dips on the upper part of this layer make it possible to place a new

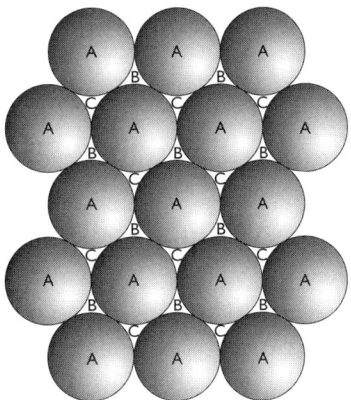

Figure 3.5
Principle for constructing lattices: hexagonal close packing (HCP) with a sequence of layers *ABAB* and so on; face-centered cubic (FCC) packing, *ABCABC*, and so on.

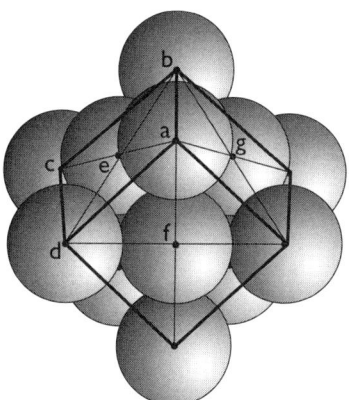

Figure 3.6
Local structure of the FCC pile. The spheres *e*, *f*, and *g* occupy the center of the cube issuing from sphere *a* and form a triangle in contact with the latter. At the center of each face like *a*, a new sphere is found, hence the designation "face-centered cubic" (note that there is no room for a sphere at the center of the cube). The axis perpendicular to the plane *efg* corresponds to the vertical plane in figure 3.5.

layer of spheres that fit on top. Each element of the new layer will be in contact with three spheres on the lower layer and form, step by step, a new, dense layer of spheres, *B*. In the process, however, only one out of every two dips in the lower layer will have been covered. Now, we add a third layer. We can do so by placing spheres in a line just above the first layer in one out of two dips on the second layer, and so on, in the pattern *ABAB*, and so on. In this manner, we achieve periodic stacking in three dimensions, known as *hexagonal close packing* (HCP). Another form of periodic stacking has the same packing fraction as the hexagonal lattice. Instead of the *ABAB* structure, one can place the spheres of a third layer, after the AB arrangement, in a line above the gaps in the first layer, A. This layer of atoms is *C*. The crystal obtained by the arrangement *ABCABC* has the symmetry of a cube. This cubic lattice is called *face-centered cubic* (FCC) packing because it has a sphere at each of the cube's corners and at the center of each face. An illustration proves helpful for recognizing this cube (figure 3.6).

Figure 3.7a and b show a central sphere on layer *B* surrounded by six spheres on the same plane and three spheres to either side on

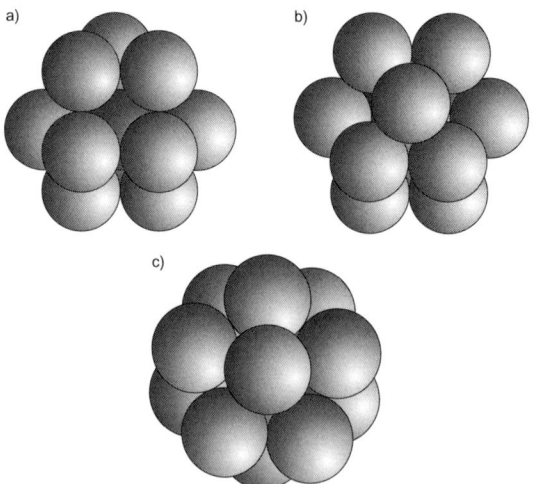

Figure 3.7

Three regular arrangements of twelve ping-pong balls around a central ball. The first, a), corresponds to hexagonal close packing; the second, b), to FCC packing; and the third, c), represents an icosahedral pile.

neighboring planes, representing the hexagonal and face-centered cubic lattices we just described. The central sphere—in fact, any sphere in a periodic pile—is in contact with twelve others: its coordination number is twelve, then. The packing fraction is a little more than 74.05%. This packing fraction is a little bit lower than what one would obtain locally for four spheres forming a tetrahedron. This difference has prompted a great deal of ink to flow. Kepler's reflections on snowflakes go on to affirm that no packing order surpasses the natural, periodic stacking of seeds in a pomegranate (FCC). Ever since, mathematicians have endeavored to confirm this conjecture. In 1998, Thomas Hales proposed a proof based on computer calculations (which he finalized a few years later). That said, mathematicians familiar with spaces on a scale far above our ordinary three dimensions aren't content to stop there. Maryna Viazovska, a young Ukrainian mathematician, recently solved the sphere-packing problem in eight and twenty-four dimensions! Most remarkably, packing spheres in n-dimensions has led to concrete applications: it provides the basis for error correction codes in telecommunications, allowing distorted data to be recovered in the event of poor transmission.

There are other periodic packings in space that are less compact than those described up to this point. Their study is the object of crystallography. Most of them cannot be obtained by placing spheres in contact. For now, let us return to periodic packings that occur in particles in suspension. In this instance, clear separations between neighboring spheres are not ensured by contact, but by interactions at a distance that keep particles at a remove from each other, with a tendency to form periodic modes of organization. This is the domain of *colloidal crystals*.*

COLLOIDS AND OPALS

The two first chapters acquainted us with colloidal particles of gold. We saw that their size accounts for color. We also saw the forms of fractal aggregates that they yield. Another class of colloids that can be produced chemically through organic synthesis in liquid phase is "latex" spheres. Monodisperse spheres of polymer (polystyrene) in solution are commercially available; the diameter varies, on the basis of conditions of synthesis, between 0.1 μm and 1 μm. In turn, they are diluted in a solution with

Figure 3.8
Photograph of the surface of a colloidal crystal composed of latex grains.

an optical index of refraction close to that of the spheres themselves; this makes the medium nearly transparent. When, after a few moments, the concentration of spheres is sufficient (say, at a volume concentration of 50%), the solution displays a remarkable rainbow effect, like opals (discussion follows), which changes as a function of the angle of observation. The particles are arranged in a crystal lattice (figure 3.8) and the light that reaches the "mega-atoms" of this "crystal" is reflected selectively, according to an angle that depends on the value of the wavelength. This phenomenon of selective diffraction is the result of phases matching between waves as reflected by atomic layers of the crystal. If this condition—a function of wavelength—is not attained, no light is reflected. The phenomenon represents the analog of *Bragg's law* for X-rays reflecting on atoms, a discovery that made it possible to study crystalline matter precisely and quantitatively. For X-rays, wavelength and interatomic distance lie on the order of an angstrom (1 Å = one tenth of a billionth of a meter). In the case at hand, the distance between planes of spheres belongs to the order of 1000 Å, that is, the order of visible light. The wavelength of light, then, serves as a gauge: a ruler where the line spacing is fitted to the distance between crystalline planes.

The crystalline organization of colloidal particles like latex spheres in solution comes from the repulsion between particles with the same charge. The spheres repel each other, yet at the same time, the vessel containing them doesn't allow them to escape; ultimately, this process yields a periodic arrangement. The mineral salts in the liquid where the spheres are floating dissociate into ions, and their charges screen the charges conveyed by the spheres, which limits electrostatic repulsion. If the salt concentration is raised, the forces between the suspended grains are no longer enough to produce crystal. In this case, the spheres remain distributed in a disordered amorphous state.

By varying the concentration of particles in solution, colloidal suspensions may serve as models to understand changing states of matter. In such an experiment, the repulsive interactions between spheres are geometrical in nature: the spheres are not smooth, or "bald." Instead, they are "hairy" and covered with a fine layer (10 nm) of linear molecules of polymers, with one end attached to the sphere (like unused tennis balls). The compression their "hair" experiences when two spheres draw near to each other gives rise to a force of repulsion between them. At weak concentrations, the spheres remain in a disordered state relative to each other. The suspension looks milky because the isolated grains diffuse the light (remember the case of milk itself, in chapter 1). At medium concentrations, between 40 and 44%, a suspension left to rest for long enough will separate into two phases: a crystallized portion on top, and an amorphous zone underneath; if the concentration of spheres in solution is raised between 40 and 44%, the crystalline fraction will increase from 0 to 100%. This process is analogous to transition between a liquid and a solid state. At the melting point, a range exists where both phases are present: a lower-limit state (the first little crystals in the liquid) at the start of crystallization and an upper limit state, when a final drop of liquid is in equilibrium with the crystal.

So what happens at concentrations higher than 50%? There is still a crystalline state, but the little crystals that form next to each other point in all directions; because they are "jammed," they cannot reorient themselves to form a larger crystal. The more the concentration grows, the more this effect increases, until a maximum concentration of 55% is reached, where disorder occurs on the tiny scale of distance between spheres. A glass has been made.

Another instance of crystalline order among microscopic objects—this time in the natural world—concerns *opals*.* Opals consist of spheres of silica sharing radii of equal length (on the order of thousands of Å). Here, it is not electrostatic interactions between spheres to produce the crystalline structure, but contacts between them. In water progressively enriched with silica, the precipitation of silica yields a gel made up of particles a few hundreds of an angstrom in diameter. As they agglomerate, these particles yield well-calibrated, small, and spherical grains that settle very slowly in the gel and form a compact periodic lattice under the effect of their weight. The beauty of jewelry made with these objects comes from the play of colors caused by the refraction of light, which occurs through the organization of parallel and equidistant planes of the silica lattice.

Metamaterials   The example of opals shows that it's possible to transpose, into similar arrangements, some properties present in atomic systems: the light visible in an opal corresponds to the selective reflection of X-rays (Bragg's law). We speak of *photonic crystals* when describing structures whose optical properties can be controlled by working with the organization of these mesoscopic constructions. The phenomenon can involve colloidal particles self-assembling in natural arrangements (figure 3.8)—or result from constructing arrangements on the scale of microns layer by layer, by means of 3D printing.

These metamaterials possess numerous other properties that are tuned as a function of the size and geometry of piles. Many practical uses are in development. We provide an example from the field of acoustics in figure 5.7.

DISORDER IN GRANULAR MATERIALS

Yet again, solid spheres will help us to understand nonperiodic structures. As for the crystalline state, our point of departure is the organization found on a single layer of the material in question. Our observations in this context will acquaint us with disorder exhibited by volume stacking.

DEFECTS IN A LAYER OF SPHERES ...

A flat pile rarely possesses the ideal structure of equilateral triangles assembled in the manner illustrated in figure 3.3a. Let spheres roll down

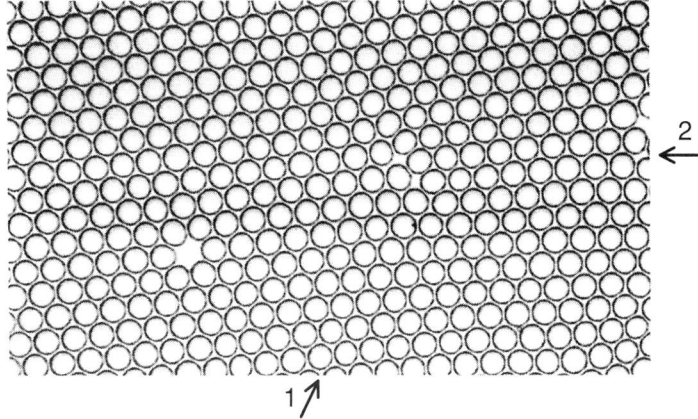

Figure 3.9

Periodic lattice of bubbles formed on the surface of water. Although the lattice may seem perfect, it still contains dislocations. They are visible by observing at a grazing angle along direction 1. The dislocation corresponds to where a row of spheres has been added in the upper part, up to level 2. A point defect (lacuna) is also evident, but it does not compromise the large-scale order.

an inclined rectangular plateau. The resulting pile will need to adjust in response to edges at a right angle. Right angles do not occur in triangular stacking—hence the local advantage of the cubic lattice. Far from the angular corner, a triangular lattice will form, but with faults needed for adjustments.

Figure 3.9 presents a layer of bubbles of the same diameter on the surface of water. If we consider the triangular arrangement, which seems to contain no defects, along the oblique line indicated by arrow 1 we can see that a line of bubbles has stopped at the point of intersection marked by the other arrow, 2. This basic stacking defect is a *dislocation*.* Large-scale order is missing; to join two points on either side of this line, one cannot rely on the whole numbers, n_1 and n_2, used to determine the vertex of a tile in the triangular lattice. Lines of *dislocations** form at the border of monocrystals and are called *grain boundaries* (as depicted in figures 3.10 and 3.11).

These images allow us to understand an important concept in mechanics, *plasticity*.* By an irreversible shift of grain boundaries, solid material undergoes deformation. A layer of bubbles or spheres that has been

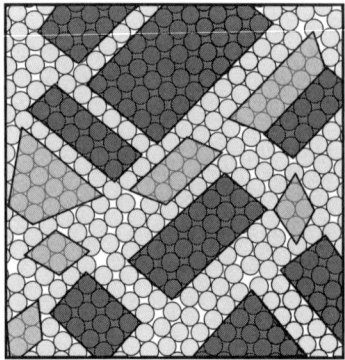

Figure 3.10
A model polycrystal appearing in a pile of quasi-monodispersed grains. Zones of triangular arrangement (light gray) are evident, as are zones of rectangular arrangement (dark gray). The grain boundaries marking the limit between zones can be viewed as dislocation lines.

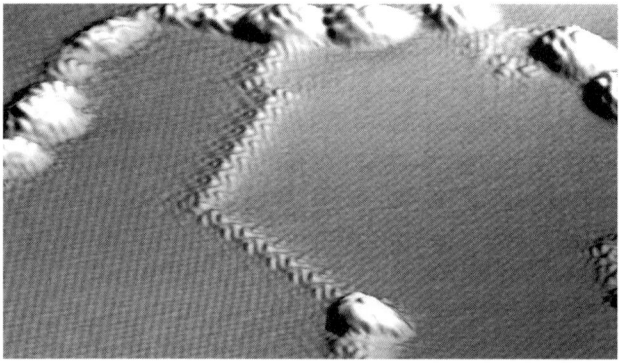

Figure 3.11
Grain boundary between two nanocrystals of graphene deposited on a copper crystal, viewed through a scanning tunneling microscope.

deformed by moving its boundaries will assume a new form of organization and find a new equilibrium without having to return to its initial position. This state contrasts with what would have occurred if only
a small force had been exerted. In the latter case, the grain boundaries
would not have moved; the bubbles/spheres would have been only a little
deformed, and they would have returned to their original form, once the
force was removed. This is the domain of *elasticity*. This elastic region may
be extended by hindering the motion of grain boundaries—which is the
function of atomic impurities in a metal alloy (for instance, carbon in the
manufacture of cast iron).

... AND THE RESULTING DISORDER

Where does the accumulation of defects, or the decreasing elementary
size of tiny crystals—the "puzzle pieces" of a polycrystal—stop? Does this
occur at the atomic level or at the level of tiny crystals consisting of a few
atoms?

Using packing models of two-dimensional discs, it is possible to reproduce the equivalent of these amorphous structures. All it takes is combining discs that do not have exactly the same diameter. This is what
has been done in figure 3.12, where a small percentage of large discs has

Figure 3.12

All it takes to obtain a disordered pile with coins of the same diameter is adding a small
fraction of bigger coins.

been added to an ensemble of discs with a smaller diameter. Voids appear, because six small discs in contact with a larger disk cannot all be in contact with each other. This local defect disrupts the regular growth of the triangular crystal and occasions an amorphous structure locally.

DISORDER IN A BAG OF SPHERES

An Icosahedron of Spheres and a Soccer Ball   When we began examining bulk packings, we affirmed that two approaches exist. We steered the first course when we obtained 3D crystals by piling up successive 2D crystalline layers. The second approach involved considering the many ways to obtain a dense packing locally starting with a single sphere, and then continuing until a packing on a large scale results. We have already considered, with cubic and hexagonal crystals, two ways of placing twelve spheres in contact with a central sphere (figure 3.6). In fact, this operation may be performed in an infinite variety of ways: two possibilities out of an infinite number—that's not much! This is all to say that one never encounters such orderly structures, unless some mechanism has been used to induce the crystalline state by force, as in the case of opals.

In fact, one possibility for placing twelve spheres around a central sphere proves to be more probable than all others. Let's start with the elementary structure of the regular tetrahedron of four balls, where one ball is resting on the three others. Forget the spheres for a moment, and try to place tetrahedra made from four spheres in contact side by side around a shared edge. Five tetrahedra can be arranged in this manner, not four or six, and even so a small void remains. Every crystallographer knows that a structure with five sides is incompatible with the crystalline state. But starting with these five tetrahedra around a shared vertex, it is possible to place another layer of ten tetrahedra, followed by five more in staggered arrangement, then, finally, a sphere at the top. If the spheres at the vertices are removed, we see twelve tetrahedra surrounding the central sphere, although a small void remains. By redistributing this space in uniform manner among the twelve peripheral spheres, the latter come to occupy the vertices of a regular polyhedron with twenty flat, triangular faces: an *icosahedron*.

This structure is most probable from the point of view of local organization because it results from the elementary structure of a tetrahedron.

What's more, it occurs in small aggregates that can be produced by projecting a few atoms onto a chilled surface. It is also found in many examples in nature, for instance the heads of certain viruses. Finally, it is evident in soccer balls (see figure 11.1), which can be pictured as an icosahedron whose twelve vertices have been smoothed over and replaced with twelve pentagonal facets to get a roughly spherical shape. Accidentally, carbon smoke has been found to contain regular molecular structures made up of sixty atoms with the symmetry of a soccer ball with twelve pentagons— sixty is the number of vertices in the soccer ball, you can check for yourself! The name for such organizations of molecules, *fullerene*, honors the architect Buckminster Fuller, who built a geodesic dome for the 1967 International and Universal Exposition in Montreal with an arrangement of cells similar to a gigantic molecule of C_{60}. Decades later, this example of the geometrical organization of carbon atoms earned its discoverers the Nobel Prize. Its local structure is the same as that of graphene, presented in figure 3.2, and also the object of numerous applications.

Creating Disorder in Volume   An icosahedron of spheres, then, forms a highly probable local structure, and therefore one that is very stable. But if, in turn, one adds spheres to the icosahedron in order to make this structure larger, the existence of voids around each added sphere leads to a diminution of compactness: incompatibility holds between a structure that is compact on a macroscopic scale and the locally organized structure of the icosahedron. The disorder represented by the void subsisting in this structure grows as further spheres are added; the structure obtained loses its "memory" of the initial germ. If a few atoms of argon are placed on a cold surface in such a way that they remain immobile, they will gather into an icosahedral structure. But if their number is increased, the structure does not persist, and they will reorganize into a crystalline form. In this case, we've moved from the domain of molecular physics to that of solid physics.

A great deal of analog and computer experiments have explored the organization of disordered piles of spheres of the same diameter; they have been conducted in England in particular. First and foremost stands research conducted by John Bernal, who before World War II had described the structure of a liquid as follows: "homogeneous, coherent and essentially irregular assemblages of molecules containing no crystal-line regions nor, in their low temperature form, holes large enough to admit another molecule." This definition led Bernal as a matter of course to the disordered sphere packing obtained by pouring the spheres into a rubber bag that was then shaken. A next step was to pour paint into the bag. Once the bag was drained, the paint that remained between grains through capillary action dried, thereby ensuring the mechanical strength of the packed spheres. Bernal took apart the structure piece by piece, scru-pulously noting the coordinates of each sphere's center before removing it. In this way, he obtained the local organizations of grains and contacts between them.

Even though disordered packings of hard spheres (like those that Ber-nal constructed) are often called random, this is not entirely the case, because exclusion effects (the fact that two neighboring spheres cannot overlap) prevent certain configurations and therefore produce a certain *short-range order** between the positions of spheres. The *Voronoï diagram** in figure 3.13a for packing on a plane still provides a powerful tool for identifying local volume structure at the level of a grain and its imme-diate neighbors. In space, the bisector planes of lines that meet in the centers of spheres replace the medians in the drawing; in this way, one obtains an ensemble of polyhedra resembling those that form when a glass bottle of beer is shaken (figure 3.13b), instead of an ensemble of polygons in two dimensions.

Describing packings of spheres based on the model of an assemblage of polyhedra is nothing new. In 1727, the English clergyman Stephen Hales published the results of an experiment he had conducted on the absorp-tion of water by peas piled in a pot and kept under pressure to preserve a constant volume: as they expanded, the peas filled the voids as they flattened and took the form of polyhedra, which he described as "pretty regular Dodecahedra" (regular polyhedra with twelve pentagonal faces). Applying the Voronoï diagram to a compact crystalline pile leads to regu-lar polyhedra with twelve faces. In contrast, disordered stacking involves an ensemble of irregular polyhedra that differ both in terms of topology and shape. The defining characteristic of these polyhedra is an average number of faces slightly higher than fourteen and a predominance (near 40%) of pentagonal faces.

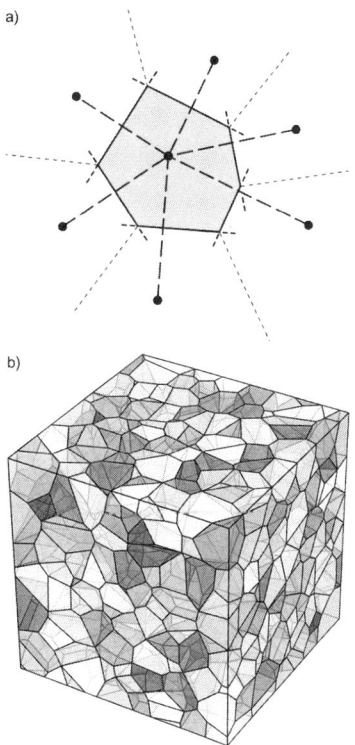

Figure 3.13

a) Voronoï construction allows a disordered ensemble of points, like the centers of the coins in figure 3.12, to be replaced by polygons in contact, starting from the medians at the centers of neighboring grains. b) This example is a Voronoï diagram in three dimensions.

PLAYING WITH SIZE

In light of the foregoing, it seems quite difficult to change the packing fraction; the next chapter will address the issue. For industrial purposes, however, it is important to vary compactness across a wide range. Often, efforts are made to eliminate voids in order to improve sturdiness and reduce the possibilities of infiltration between pores. But at other times, the point is to make lightweight materials; this is the case for certain heat-conducting materials (such as copper) that are used as heat exchangers with a fluid circulating within the porous space. So how is a granular material made more or less compact in an appreciable manner? We will consider how changes to the form of nonspherical, nondeformable particles lead to significant packing fraction variations. Another possibility we will examine, first, is making structures with grains of different sizes, which reduces empty space.

Apollonian Packing   To increase the packing fraction of a pile consisting of spheres of the same diameter, one solution, naturally, is to fill its voids with smaller grains. The "diameter" of the void inside a regular tetrahedron is equal to 0.225 times that of the spheres forming the tetrahedron (see figure 11.1). Theoretically, this operation can be performed by means of successive iterations, whereby the spaces left empty at a given level of the pile would be filled by spheres of progressively smaller diameter, down to an infinitely small scale. Apollonius of Perga envisioned this so-called *Apollonian packing** solution (represented in figure 3.14) more than three centuries before the Common Era. In passing, we should note that this pile's geometry is an illustration of a fractal object (cf. chapter 1) because it presents the same average geometry at different levels of magnification.

It's not very realistic to try to achieve this optimal packing experimentally. We can imagine placing spheres with a well-calculated range of sizes at zero gravity and, by means of stretching and squeezing, filling the assigned space perfectly. Combining grains of various diameters allows us to approach the ideal. Thus, we can start with an assortment of

Figure 3.14
In this simulated Apollonian model, the spheres fill a space whose porosity tends toward zero, as a function of the size of the smallest spheres.

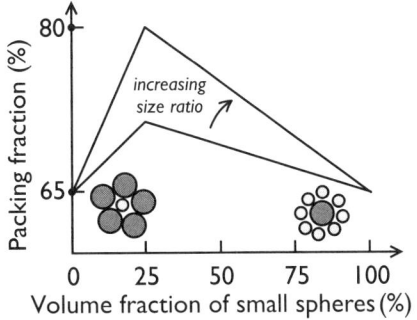

Figure 3.15

The packing fraction of a binary mixture of spheres presents a maximum value for a fraction of small spheres of about 25%. This maximum value increases as the ratio of radii grows.

spheres with the radii R_1 and R_2 close enough in size that the smaller ones won't move freely in the gaps of the larger ones, say, $R_2 / R_1 = 2$. Hereby, we observe (figure 3.15) that, whatever the relative percentages of these two kinds of sphere may be, the packing fraction is higher than if we only had spheres of a single radius (needless to say, the packing fraction will be identical for 100% of spheres in category 1 and 100% of spheres in category 2). A tiny percentage of the smaller spheres will fill out the gaps between the large spheres. Conversely, if there are few large spheres, their full volume will replace a volume that, if occupied by small spheres, would have included voids. Starting out from this binary combination, we can envision ternary combinations, too—each time a new category of spheres is added, the packing fraction C increases.

This chapter has acquainted us with diverse modes of organization of granular media, starting with the simple packings of spheres and discs. We began with periodic piles and then moved on to the more common phenomenon of disordered piles. This disorder will be our guide for the remainder of this book. The geometry of individual grains and their size distribution, as well as the way they assemble into piles or flow, will be described in terms of averaged quantities.

4

THE PACKING FRACTION

Good measure, pressed down, shaken together, and running over will be given to you.
Gospel of Luke

The preceding chapter familiarized us with various types of packings, ordered or not, and presented their geometric properties. Now we will address more fully how the *packing fraction* of granular materials—especially disordered ones—may be adjusted.

THE LIMITS OF PACKING FRACTION

In any given material, small-scale order or disorder leads to highly variable average properties. These properties are of interest to anyone wishing to model the macroscopic behavior of a granular material without needing to know all its details.

One of these parameters, introduced in chapter 3, is the fraction of volume that grains occupy—in other words, the packing fraction. As we have seen, it depends on the way the grains are poured or assembled.

SEVERAL HIGHER VALUES

Drop, one by one, metal spheres of the same diameter into a container. If the vessel is filled with a viscous liquid, their falling speed slows down; as soon as the spheres come into contact with other spheres already at rest, they assume a stable position. In this case, the packing fraction is on the order of 60%. If one pours the spheres into a rigid receptacle in the absence of liquid but all at once, a lower value may result—on the order of 56 to 60%—but this process is poorly reproducible. To pack spheres compactly, on the other hand, the container should be shaken and vibrated as it is being filled; alternatively, one may drop the grains in small quantities from a relatively elevated position, so that their shocks result in local reorganization. Now, the packing fraction reaches almost 64%. This is the maximum value for a disordered pile of undeformable, identical spheres.

What happens in the interval between this value, 64%, and the 74% obtained in the case of the most compact ordered packing we encountered in chapter 3? If a value between these two figures is attained, the system is probably partially crystallized. All this attention to limit values is far from idle. A change of state has taken place, which is analogous to a *phase transition** with an increased order—just like the one observed when a liquid becomes a solid at its temperature of solidification.

OTHER REASONS FOR VOIDS

Varying compacity is not limited to granular materials. All materials (except water) increase in volume when the distance between atoms or molecules increases in response to rising temperature or diminishing pressure. Variations of the packing fraction in a granular medium also depend on the current state of the pile, which, for its part, results from whatever deformations it has previously undergone. Packing, then, does not characterize the material so much as it represents a signature of its internal state. It is impossible to speak of the *absolute* packing fraction of salt in a shaker, soil in a desert, or snow on a mountain. It is more accurate to speak of the material's state of packing. All the same, however, one may ask whether well-defined packing states exist with values that characterize one material in particular, not just values resulting from the packing condition.

If one drops spheres into a vessel, some stop as soon as they make contact with the pile in its the process of accumulating; others roll until they stabilize by coming to rest on three other spheres. The piled state formed in this manner reflects an overall balance, since each sphere is in a state of equilibrium under the effect of the forces exerted by those that surround it: the packing fraction doesn't enter the equation as a cause or criterion of the pile's stability: contacts and the balance of forces are what count. If, after filling, the packing fraction has not achieved its maximum value of 64%, that value can be increased by slightly shaking the ensemble. However, it can also be reduced, if too much shaking occurs. What a headache! Increasing the packing fraction requires, first, that grains move collectively. In this instance, what limits its value is not the individual equilibrium of each sphere (the spheres have already reached an equilibrium) so much as the fact that the spheres cannot overlap; *steric exclusions* are at work.

Accordingly, a given granular material has two packing fraction limits: a maximum value related to the fact that two spheres cannot occupy the same space, and a minimum value tied to the equilibrium of each sphere, which requires a minimum number of contacts to stay in place. These limit values depend on the nature of the material: the shape of grains, their size distribution, the friction that occurs between them, as well as, in some cases, the effects of adhesion. For spheres of the same size with a significant amount of friction between grains, the minimum value lies at $C = 0.56$ whereas the maximum is 0.64. We will see how these limits also depend on grain shapes.

THE HOLY GRAIL: A BAG OF MARBLES

One problem that has received theoretical attention in recent years involves geometrically achieving the most compact stacking possible. There are many ways to arrange spheres alongside each other. In fact, numerical evaluation of the possibilities for assembling a discrete number of spheres (say, $N = 50$) to produce an optimally compact filling of disordered elements defies even the most powerful computer. The undertaking belongs to the class of what is known as "complex" or "non-polynomial" problems, because the number of operations necessary, as the number of spheres increases, grows faster than any power of N. Fortunately, approximate solutions exist. In this spirit, remember an image related to our hypothetical bag of spheres. Say we want to improve the compactness of a given filling. It would be possible to make a series of small adjustments, but this method wouldn't work too well. A better solution implies a significant reorganization on a global level. To this end, one must pass through less compact states in order to have a chance at reaching better solutions later on. This is a problem confronting optimization efforts in general. As previously noted, the trick is to shake and stack. If the pile consisted of sugar cubes that have been roughly poured into a sugar bowl, there wouldn't be much to gain by compressing them ... unless the point is to make powdered sugar! But by shaking gently and allowing gravity to lead the grains toward a more favorable overall configuration, a much better result is obtained.

In practical terms, the parameter governing the process of compaction by vibration is the relationship between the amplitude of the grains' oscillation and the acceleration of gravity—that is, the effect of their weight. If vibration is too strong, decompaction wins out: it has the effect of keeping the material in a loose state. But if, on the contrary, vibration is weak, it simply plays a minor destabilizing role, which enables the grains to achieve a denser state.

What happens if a granular material is subjected to continuous vibration? This occurs every time you fill a jar with powder or grains and shake it so it will become denser. Compaction is ongoing, but at a rate that decreases constantly. At first, the material is loose, and it's easier for the grains to reach a denser state. But as the packing fraction gradually rises, they need more and more time to reach a denser state. In other words, the rate of compaction is proportional to the difference between the current value of the packing fraction and its maximum value. In mathematical terms, this property leads to a logarithmic increase of compactness as a function of time: the increase of compaction between 10 and 100 seconds is the same as between 1,000 and 10,000 seconds. This evolution is a synonym for *slow dynamics,* which one also finds in the physics of glass, in the sense of a material between solid and liquid, which has not achieved a state of equilibrium.

Slow compaction under the effect of vibration has inspired an array of physical models, including the "random parking" model. Start by considering how a parking structure fills up with cars, one by one. If a large area is available and no spots are marked, each car will find a place quickly. But the fuller the parking lot is, the greater the time required for finding a spot. In consequence, the number of spots occupied grows as the logarithm of time (the inverse of exponential).

MINIMUM COMPACTNESS

Increasing Porosity   Minimum compactness poses even more problems! Theoretically, and in computer simulations, it is possible to proceed efficiently by randomly removing grains from within the medium. When a grain is taken out, an open space is created, which can be occupied by a neighboring grain. Removing certain grains, however, does not necessarily disturb the overall equilibrium, and the pile remains stable (albeit at a slightly reduced compactness); it's like a game of pick-up sticks, with spheres instead of sticks. If one continues to take out grains, collective rearrangements will occur from time to time, in the course of which the packing fraction suddenly increases. On the whole, however, the packing fraction diminishes, and the material tends toward its lower limit. The process may be likened to what happens in rockfill dams, where internal erosion leads to the progressive disappearance of small grains (which the large ones cannot hold back), yielding a loose state subject to collapse.

One can try to increase the voids in a material to make it lighter, to attenuate the sound that passes through it, or to enhance its thermal insulating properties. Using spherical grains has a limited potential for obtaining a loose packing since these particles are apt to roll and reorganize themselves more compactly. To restrict their mobility, one can focus on the individual form of the grains or on their surface state. Evenness, roughness, stickiness, or moisture can favor the emergence of additional voids and vaults (which we will discuss in chapter 5). Specific procedures make it possible to ensure a reproducible minimum packing fraction. In soil science, for example, a minimum is defined as the result of simply dumping material into a container from a small height.

In describing these borderline cases, we have not taken into account the *cohesion** that can exist between grains. The presence of bonding forces between grains influences the equilibrium conditions and, therefore, their minimum packing fraction. Acrobats can form a more rigid human pyramid if they hold each other firmly instead of simply leaning against each other. Likewise, a grain can remain in a state of equilibrium only when it rests on one of two other grains. Without getting into the details of these interactions, we can see that if the latter are strong enough to block a contact between two grains irreversibly, it is possible to achieve structures that are very loose and, at the same time, rigid in mechanical terms. For instance, the fine powders used in printing inks consist of lightweight grains that interact through forces of electrical origin, so-called *van der Waals force,** which yield a packing fraction as low as 0.3 while giving the powder a downy texture.

Model Sols and Gels   An experiment you can perform at home—called the Cheerios effect—provides a model for the compaction of grains in two dimensions. As the name indicates, this is done by pouring the breakfast cereal into a bowl of milk. The grains of the suspension gradually draw together and stick to each other under the influence of the surface, which achieves a state of balance between their weight, Archimedes' buoyancy, and the surface tension that bends the liquid around each grain. It's comparable to when a mattress in a big bed is too soft and two people are drawn together … by purely mechanical deformation! Laboratory research using spheres of wax floating in water has made it possible to devise a model of the formation, over time, of aggregates with a fractal geometry (figure 4.1)—the same kind as the colloidal aggregate we saw in chapter 2 (figure 2.9). As they take shape, the clusters move freely in solution. The *viscosity** of the suspension (the colloidal solution is called a *sol**) then increases with the size of the clusters and their crowding. When the concentration of spheres exceeds a threshold value, a continuous, ramified net of grains sticking to each other is obtained, which extends from one end of the bowl to the other. Here, we have the image of a *gel,** a weak solid that, in general, is highly deformable. Fruit gelatine is one example worth tasting.

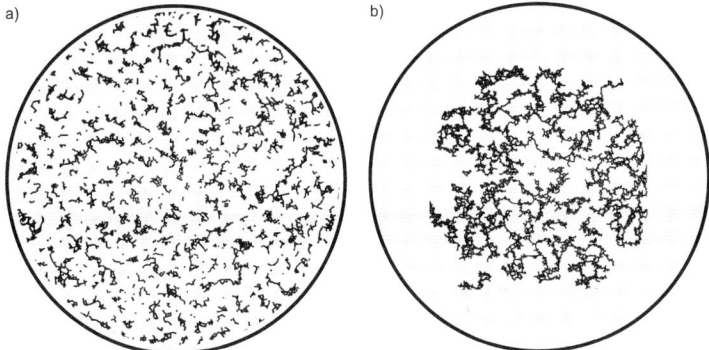

Figure 4.1

Wax spheres cast at random onto water. Under the effect of attraction due to surface deformation, they are drawn together and attach to each other. The images capture two points in time after the beginning of the experiment. The first image represents a) a *sol*, which at a later stage becomes a continuous solid, namely b) a *gel*.

This somewhat lengthy presentation has enabled us to picture the possible states of a granular medium in terms of packing fraction, and to specify its limit values. Once these limit values are known, one can define a *relative packing fraction* with a value ranging from 0 to 1 as the packing fraction varies between its minimum and its maximum value. This compactness provides the packing state in relation to limits that depend only on the material and govern mechanical properties.

PARTICLE SHAPE EFFECTS

Up to this point, our attention has been limited to simple objects such as spheres. But the vast majority of grains of vegetal or mineral origin have more complex shapes, and the sphere model is poorly suited to them. At present, major research is underway to develop specific grain shapes whose packing will be controllable, with an eye toward practical applications such as in architecture (figure 4.2) or landscape preservation. The interest in these particles under development is evident when they're poured into a pile on a surface. These piles are quite stable, but with limit angles larger than those of rounded grains: they also prove more resistant to avalanche effects.

Figure 4.2
This experimental architectural structure shows how a rigid wall may be obtained with a packing that has very little density.

HIGHLY ELONGATED GRAINS

Let's turn to elongated objects: all it takes is a bag of rice to see that grains arrange themselves in parallel locally, in order to fill the space better. Well-arranged parallel pencils—to take an extreme case—can occupy 91% of space, as in a 2D packing of disks. But if we drop these same pencils at random, they will overlap like pieces in a game of pick-up sticks: because of the friction between them, their packing fraction remains low. By incorporating metallic fibers into a resin matrix that is then polymerized, the resin is mechanically reinforced. If the concentration is high enough that the (conductive) fibers start to tangle up, a continuous electrical path results. Adding fibers to high-performance concrete (which we will present in chapter 10) has made it possible to build structures that are extremely lightweight and elastic, such as the Seoul Peace Walkway, whose bridge deck is only a few centimeters thick.

It is possible, then, for the same concentration of fibers to yield two highly distinct states—as different as a disordered tangle of fibers and a structure of aligned fibers. If we simply drop the fibers in disorder, we get the first type of structure. But forcing a solution containing fibers through a funnel or a nozzle so they converge encourages an aligned state. This is what manufacturers of polymer fibers try to do: the mechanical qualities of Kevlar, a synthetic material, derive from alignment through extrusion of the elongated molecules that compose it.

Long fibers of this sort are quite interesting, but following everywhere they lead would take us too far away from the grains at issue here. Let us be content, then, to stress that the varying states of compactness of these piles, depending on filling conditions, is infinitely larger than those of the spheroid objects to be discussed in the following pages.

M&M CANDIES

As strange as it seems, the role that the shape of particles plays with respect to assemblage properties of grains "in bulk" was little known and under-studied for a long time. (Perhaps researchers had too much confidence in the marble-ball model!) In 2004, experimental results published in the well-respected journal *Science* showed that the sphere is not the optimal rounded shape for achieving a compact disordered assemblage. Research-ers who filled jars with M&Ms observed that their elongated and flat-tened form is more efficient (figure 4.3). Whereas a disordered assemblage

Figure 4.3
Whereas a compact disordered packing of spheres with the same diameter is limited to a maximum packing fraction of 64 percent, M&M candies reach a significantly higher value, 71 percent.

of spheres of the same size has a packing fraction of C = 64%, M&Ms can reach 71%. By modifying the elongation of these ellipsoids, the density of assemblies can be made to reach a maximum close to 74% for a ratio of 3 to 2 between the largest dimension and the smallest dimension of the elipsoids. The candies' shape allows them to optimize their arrangement beyond what spheres could achieve. Thus, where spheres present six contacts, M&Ms can have eleven! Beyond this maximum value, the packing fraction of the assemblage decreases progressively as elongation grows. This is because of an excluded-volume effect that occurs with highly elongated objects, leading to bigger and bigger voids, which grains are unable to fill.

SIMULATED PACKING

More recently, researchers have been able to generalize this type of behavior for various shapes of grains in two and three dimensions. In this collaborative endeavor, each group has numerically simulated a deposit of grains of different shapes, for which common parameters are defined as angularity, elongation, and tortuous surface shape. These parameters happen to be less important for packing fraction than the grain shape's deviation from sphericity. To this end, one draws spheres (or, in two dimensions, circles), inscribed and circumscribed to radii R_1 and R_2 as depicted in figure 4.4a. The parameter $s = 1 - R_1 / R_2$ is an index of *asphericity*; in the case of a perfect sphere, it is equal to 0, and it grows in proportion to an object's deviation from this ideal. The packing fraction increases as a function of this index for all the shapes at issue, and it reaches a maximum value lying around for s = 25%. This means that the high compacity packing fraction observed for M&Ms is not specific to their particular shape; it represents a general tendency that can be observed in all grains that depart slightly from a spherical formshape.

INTERLOCKING GRAINS

Certain grain shapes have a quite peculiar property: they interlock, get in each other's way, and become entangled without bonding. This is the case for Lego blocks, puzzle pieces, interlocking paving stones placed

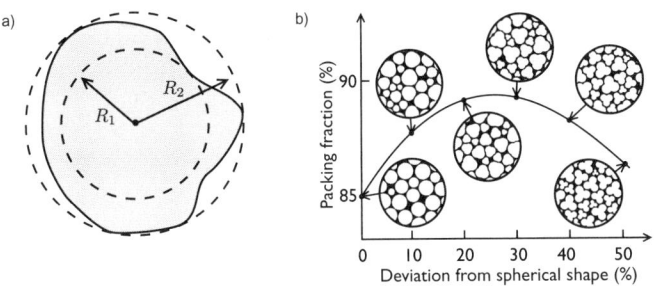

Figure 4.4

a) Measurement of deviation from circularity; b) packing fraction (C) as a function of deviation from sphericity (for packings of particles with the same size distribution). Note that the ideal shape for making a compact pile is obtained when deviation from sphericity lies at about 25%.

directly on sand without a cement bed—or paperclips in a box (which always come out *en masse*, even when you want to pull only one out!). All these shapes are nonconvex: their surfaces form hollows and bumps. One important effect of this property is the capacity to make several points of contact, as seen on figure 4.4.

Aggregates represent an important class of nonconvex objects with key roles in many sectors of industry. Couscous (figure 2.8) is one example of an aggregate. Metallic powders are often made up of aggregates of crystallites, that is, packings of small monocrystals. Recent studies have shown that where the packing fraction is concerned, these grains obey the same rules as the nonspherical grains previously discussed: they can form piles of higher compactness than those made with spheres.

Today, it is possible to manufacture grains in complex forms: Z- and U-shapes, stars, particles with hooks, and more. The stability of the packings they form is a matter of great interest. Shaking them encourages them to become entangled. This fact illustrates that the resistance of granular materials does not necessarily depend on their packing fraction. Indeed, such piles of special "grains" display a low packing fraction, close to that of disordered piles of long fibers. Their exceptional mechanical resistance without a bonding agent results from the contacts between particles and forces of friction at play.

The efficiency of complex grain shapes (e.g., crosses) for inhibiting movement can be determined by considering the stability of piles obtained by pouring them out. Not only will their repose angle be much higher than in the case of rounded grains, but these piles will also resist small-scale disturbances without causing an avalanche.

ANGULAR GRAINS

The stones employed for construction, obtained by exploding hard rocks in quarries, generally present angular edges. These grains are used in foundations and rockfill dams, and as bedrock for railways (the ballast discussed in chapter 2); they also find application in paving roads (asphalt, which is obtained by mixing them with bitumen) and concrete (when mixed with cement). Packings of angular grains usually display a lower packing fraction than piles of spheroids while generally proving to be more resistant. Whereas, between two spheres, only one point of contact is possible, two grains with flat surfaces can be in contact at multiple points and form stable structures; one brick will hold to another brick, but a rounded grain will hold to another rounded grain only if it has contacts with other grains!

An extreme case concerns ancient monuments made from blocks of carved and fitted stone, which ensure a high level of coherence because of their weight and resistance to movement without bonding. When the shapes are regular, it is possible to obtain piles with maximum internal surface contact and maximum packing fraction. In a disordered state, the number of face-to-face contacts is low. All the same, however, computer studies have recently shown that these contacts are directly responsible for the mechanical strength of assemblies of polyhedra. In chapter 6, we will see how face-to-face contacts constitute zones where forces are concentrated. Figure 4.5 shows a computer simulation of a pile of polyhedral grains with different numbers of sides. Simulations of this kind clearly show the non-monotonic variation of packing fraction as a function of the number of faces and concentration of forces around face-to-face contacts. The effects of an aspherical shape can be observed even for polyhedral grains with more than five hundred sides, which are practically spheres!

Figure 4.5
Examples of packings of irregular polyhedra with eight, twenty, and forty-six faces produced by numerical simulation.

CRITICAL PACKING FRACTION

DEFORMATION OF GRANULAR MATERIALS

Mechanics concerns not just states of equilibrium in systems, but also motion and slow deformation within them. Such information is important to consider when constructing buildings (remember the Tower of Pisa!), as well as in relation to the flows of granular materials we will discuss further on. There are different ways of deforming a granular material. The simple, horizontal motion back and forth of grains in a bowl produces displacements of elementary layers, which slide over each other. This is *shearing*, as mentioned earlier. The word calls to mind the two blades of a scissors moving alongside each other to cut a sheet of paper.

Another mode of deformation is *uniaxial compression,** that is, motion that occurs along a single axis. This widespread operation leads to the reduction of height in a pile contained by a cylindrical vessel. Instead of applying pressure in just one direction, one can also compress grains in several directions at once. This occurs, for instance, if the grains are contained in a flexible membrane like a rubber balloon put in depression: this process is called *isotropic compression*. In this particular example, the assembly solidifies.

In order to produce these deformations or *strains*, which consist of changes in form for the ensemble of granular material as a whole, force must be applied to the surfaces limiting it; in this context, one speaks of *stress.** The relation between an applied stress and a deformation underlies solid mechanics. This relation is linear when we pull on a rubber band

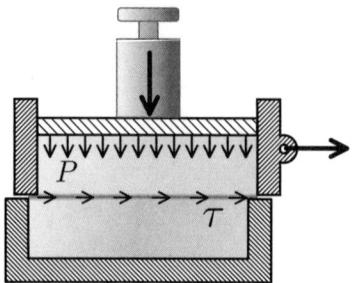

Figure 4.6

In a shear cell, the granular medium is contained by the pressure, P, between limit planes. Progressively, a tangential force (shear stress τ) is applied until rupture occurs in the two halves of the sample. The resistance to motion is the value of shear stress when abrupt motion is triggered.

to extend it, or push on a spring to compress it: in both cases the stress increases in proportion to the elongation. The ratio between stress and strain is a coefficient of elasticity, which is a material property.

The study of varying levels of packing fraction in granular materials is conducted by means of commerical devices developed for soil mechanics: *rheometers* (etymologically, "flow measurers"). They allow stresses to be applied to granular material and measure the deformations that develop. Figure 4.6 shows the example of a shear cell, a way to perform a cut by moving two flat surfaces on top of each other. Force parallel to the surfaces is applied, and the relative motion of plates is measured. So that the packing can freely change form during the shearing process, the vertical displacement of the upper wall must be left free, with pressure ensuring that grains remain in a state of containment.

CRITICAL STATE

A major observation made by means of these rheometers is that variations of the packing fraction depend on its initial value (figure 4.7). If the packing is loose at the outset, the material will contract in the course of shearing, and its compactness will increase. Conversely, when packing is dense at the start, the material will dilate, and its packing fraction will diminish. In either case, after sufficient shear deformation, the same steady value is reached. This *critical state** does not depend on the initial

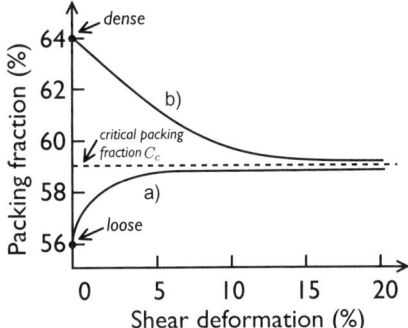

Figure 4.7
Variation of the packing fraction C of a granular packing, starting either from a) loose or b) dense packing, with continuous increase of shear deformation.

value of the packing fraction; it is purely a function of the granular matter's nature. For a packing of identical undeformable spheres, the *critical packing fraction** is $C_c = 0.59$, nearly halfway between the loosest and densest limits of 0.56 and 0.64, respectively.

In any event, granular matter only remains in this critical state if shearing continues in the same direction. If the shearing direction is changed several times in alternation, the packing fraction will continue to increase. In order to observe compaction of this kind, one need only tilt, by a few degrees, a tube containing noncompacted salt one way, and then the other. In the course of repeating the operation, one can see that the level of salt decreases after each cycle and packing fraction increases.

DILATANCY

DILATANCY AND COMPACTNESS

Anyone who has walked barefoot on the beach at low tide will have observed *dilatancy:** sand that is uniformly wet on the surface becomes dry around the foot (figure 4.8). The sand is already compact because of the effect of waves that have passed back and forth over it. The pressure of each step deforms the sand, and its compactness diminishes where the water is sucked in: at the same time, the sand dries out around the foot. An experiment one can conduct at home with dry grains will illustrate this variation of compactness. First, pour some fine-grain salt into a tube

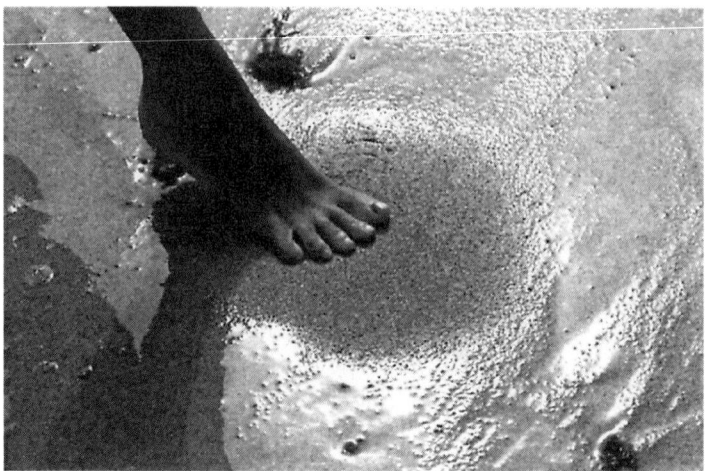

Figure 4.8
Putting a foot on the wet and compact sand at low tide leads the surrounding area to dry out. This is the phenomenon of dilatancy.

and compact it as described above. Then, with this dense state in place, tilt the tube. As the tube is slowly returned to its upright position, one can see that the salt level is higher than its initial level. The displacements of grains have led the total volume of the salt in the tube to increase, thereby increasing its porosity. This corresponds to the second type of behavior described by curve b in figure 4.7.

At the end of the nineteenth century, the English physicist Osborne Reynolds described the variation of packing fraction in response to shearing. The dilatancy is a consequence of the rearrangement of grains under the effect of shearing. In the experiment with the tilted tube, we can see that the flow starts near loose parts of the surface, that is, where a change in volume is easiest to produce; then, it spreads throughout the volume as a whole, by successive avalanches of grains of salt, in keeping with the tube's inclination. At the bottom, the sides of the tube inhibit dilatancy—and therefore the rearrangement of grains.

LIQUEFACTION

Abrupt changes to packing fraction in soil can lead to disaster—as in the case of landslides, when torrential downpour doesn't drain fast enough,

Figure 4.9
Tilted building as a result of soil liquefaction produced in an earthquake.

or when a soil is not sufficiently compact and suddenly deforms in an earthquake (figure 4.9). For just a moment, the confining pressure initially maintained by the grains is held by the liquid in the pores; in consequence, a building will sink, or an object less dense than the suspension will return to the surface. When the grains go back to where they were, the displaced objects remain at their new position. This phenomenon is called *liquefaction*. It's a bit like what happens when one walks close to the edge of the waves and the surface suddenly turns into a liquid. The situation differs from that of dry granular media, where the layers of grains resist shearing deformations because of friction between them; in contrast, liquids (which contain grains in suspension) resist only isotropic pressure.

Without wholly abandoning the trusty bag of spheres, which helps us understand the significance of small modifications in packing for a granular pile's properties, current research is examining, in more general terms, the effects of grain shape and the properties of packings in order to control compactness and, ultimately, optimize structures made with these grains. Let's move on and take a look at a scree made from jumbled rocks, stone, and gravel, all of highly different shapes and sizes.

5

MAKING CONTACTS

Football is not a contact sport, it is said: it is a collision sport. Dancing is a contact sport.
Keith Jackson, American football coach

We have discussed the *positions* that grains occupy in a packing. It's also possible to study them starting from the *contacts* that exist between these grains. This grain-contact duality lies at the foundation of the mechanics of granular media. But unlike grains, which last, contacts are ephemeral, dynamic, and progressive. They are *ephemeral* because they are subject to random encounters that increase with agitation; *dynamic* because such collisions store, then release—like springs—the energy that particles acquire in displacements; and *progressive* because, upon closer examination, as long as the contact lasts the small asperities on the surface of grains become deformed and abraded under the effect of highly concentrated forces at these points.

TWO GRAINS, ONE CONTACT

When he first observed the waves that now bear his name, Heinrich Hertz experimentally confirmed James Clerk Maxwell's theory that light is an electromagnetic wave. Previously, the German scientist's doctoral work had focused on a little-studied phenomenon, what is now known as *Hertzian contact*.* Here, he had described how two spherical objects pressed against each other undergo deformation, a process that plays a fundamental role in mechanics—for example, the way a tire is deformed on contact with pavement; it also bears on sizing ball bearings, hardness testing of materials, and mapping the topology of a surface by atomic force microscopy, all of which form part of the new discipline he initiated: *contact mechanics*.

To connect applied force and deformation at the level of a contact, one must also consider the physico-chemical nature of the interfaces at the basis of cohesive interactions; the latter may, for instance, prevent a damp sandcastle from collapsing (chapter 9). The relative importance of a contact, in mechanical and physico-chemical terms, depends on the scale of grains: for grains smaller than a micrometer, interaction proves essentially physico-chemical in nature and involves electrical, van der Waals forces. At a scale above a micrometer, the dominant interactions between grains depend on the elasticity of contact and friction between particles.

In this light, let's look at the flattening that occurs between two spherical grains sufficiently smooth that we can leave aside the presence of surface asperities. The contact area undergoes deformation. If the two grains are elastic, this deformation vanishes when the force is removed.

Hooke's Law

By applying compression, F, to the ends of an elongated object, its length, L, is reduced by a small distance, δ (figure 5.1a). For a compressed rod of section A and length δ, the relationship between F and δ is linear: F is proportional to δ: the more rigid the object, the higher the coefficient of proportionality (its stiffness). When force acts on a surface A, we use the proportionality between stress $\sigma = F / A$, and deformation $\varepsilon = \delta / L$. The linear relationship is written

$\sigma = E\varepsilon$, where E stands for Young's modulus. This is Hooke's law, named after Newton's contemporary Robert Hooke, who gave it the formulation, "ut tensio, sic vis," that is, "deformation is proportional to stress."

The proportionality between force and deformation that applies to an elastic rod does not hold for two spherical grains being pressed together. In this case, the contact area increases with applied force. It

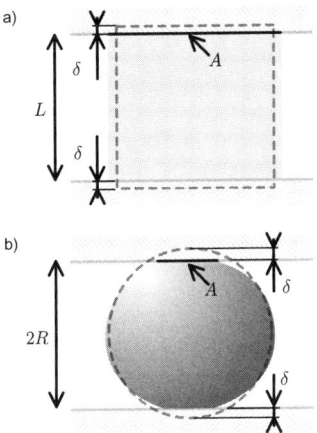

Figure 5.1

a) A cube of area A and height L experiences relative deformation, $2\delta / L$ proportional to the stress applied, or pressure, F / A. b) In the case of a sphere, deformation increases less quickly with the force, F, because the contact area also increases with pressure.

can be demonstrated that, for two spheres of radius R, or for a sphere compressed between two planes, (figure 5.1b), this contact area A varies as $R\delta$, whereby δ represents the vertical displacement of the center of the sphere, which flattens at the level of contact. Unlike the case of a rod, the pressing force increases faster than deformation (figure 5.1b). The ratio between force F and δ is expressed as the Hertz relation: $F \simeq E\sqrt{R}\delta^{3/2}$. To calculate the stress at the level of contact, F is divided by the contact area A which is proportional to the radius R. Thus, for equal forces, the smaller the grains the larger the stress at the level of contact. For two steel spheres a millimeter in size subjected to a compression force corresponding to a kilogram, the deformation is on the order of 1 μm, and the stress of 10^9 Pa, 10,000 times higher than atmospheric pressure.

What happens if force F is suddenly removed? The two grains are set free again and repel each other, like a spring being released. The elastic energy stored in the contact zone manifests itself in the speed of separation. But that's not the whole story! To investigate this process more closely, drop a "super bouncy ball" onto the ground. It will bounce back up almost to the height from which it has fallen. The potential energy lost on the way down is transformed into kinetic energy. When the ball strikes the ground, this energy briefly turns into elastic energy that deforms the ball, then back into kinetic energy … which turns into potential energy as it goes up, almost to where it started. "Almost": during the ball's excursion, part of the energy has dissipated. In order to evaluate the process, one defines a *coefficient of restitution*,* $e = v_2 / v_1$, where v_1 and v_2 stand for velocities just before and just after the ball hits the ground. For an elastic collision, e equals 1; in contrast, a wholly inelastic collision between two soft grains equals zero: all the energy is dissipated on impact, and the object does not bounce back. This coefficient equals 0.6 for glass spheres, and it can reach 0.9 in the case of polished steel.

Where does this dissipation of energy come from? For the most part, from the heat produced on impact. But there's more. We saw above that the contact concentrates stresses. Whatever the grain under examination may be, observation under a microscope will reveal surface roughness made of larger and smaller asperities that become deformed under the effect of pressure. Now, the deformation is not a matter of the grains' elasticity so much as their plasticity. Just like a thumbprint in modeling clay, a sufficiently high stress will make one sphere irreversibly mark the surface of another sphere.

In a more general sense, if the force, F, or the speed of collision v_1, is small, the deformation at contact is elastic, and the coefficient of restitution approaches 1. On the other hand, when a certain limit is exceeded, the coefficient of restitution reflects plastic deformation and takes on a weaker value, depending on the nature of the material as well as the speed of collision or force exerted between two grains.

An even stronger stress may be applied, with the consequence that mechanical contact proves fatal for the grains. Not only are the asperities damaged but also the grains themselves crack, or even break apart. Ball milling (see chapter 2) takes advantage of highly concentrated stresses when grains are progressively fragmented through repeated collisions with falling millstones.

FRICTION BETWEEN TWO GRAINS

In chapter 4, we saw how the shape of grains and the nature of their contacts influence the filling and packing fraction of a pile. How do these contacts react to force? The science of interacting surfaces, or *tribology**

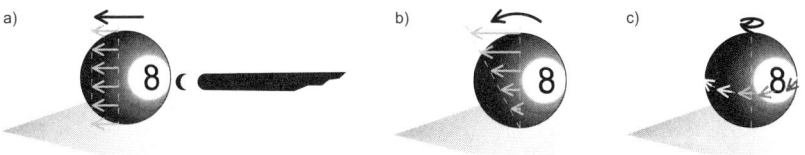

Figure 5.2
A billiard ball set into motion with a queue may a) slide; b) roll; or c) pivot depending on the direction in which it is struck.

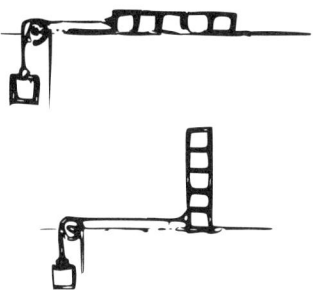

Figure 5.3
Leonardo da Vinci's drawing, reproduced here, indicates that a solid slides on a horizontal plane with the same coefficient of friction, independently of contact area.

(in Greek, *tribos* means "rubbing"), explores friction between grains subjected to an external force. When contact occurs between two grains, in addition to normal displacement (moving closer or apart), there can be three other types of relative motion at the level of contact: sliding, rolling, and pivoting (figure 5.2). These movements are what determine a ball's trajectory in a game of billiards, as we will see a bit later on.

SLIDING

In his illustrated notebooks, the codices (figure 5.3), the Renaissance genius Leonardo da Vinci devoted special attention to friction between solid surfaces on a plane, a phenomenon he needed to understand in order to invent new devices such as the ball bearing. Let's take a look at one of his many drawings. Leonardo observes the *threshold* force F_T at which a brick of mass m (and weight $F_N = mg$) placed on a horizontal

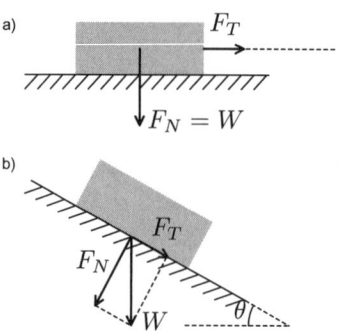

Figure 5.4

a) Geometrical construction providing the composition of normal and tangential forces, F_N (weight) and F_T (equal and opposite to the friction force) on a horizontal plane. b) Normal and tangential forces on an inclined plane.

plane and pulled will start to move. In addition, he notes that this force is the same, no matter which side the brick is placed on. In this way, a *coefficient of friction** μ_s is obtained through the ratio F_T / F_N (figure 5.4). We should note that this coefficient does not depend on the nature of materials in contact, but mostly on their surface state with imperfections and impurities at a small scale: two plates of glass perfectly processed and smooth would remain stuck to each other on the atomic level and defy efforts to slide or separate them.

The Law of Friction

Instead of placing a brick on a horizontal plane and measuring the force of friction that resists motion, one can place the brick on an inclined plane and measure the limiting angle after which it starts to slide. Figure 5.4 shows a geometrical construction that brings the two experiments together. Component $F_N = W \cos\theta$ is exerted perpendicularly to the contact surface, and the other component, $F_T = W \sin\theta$, is exerted to make the brick slide. Coulomb's law holds that the value of the ratio at which sliding first occurs, $F_T / F_N = \tan\theta_s$ gives the friction coefficient $\mu_s = \tan\theta_s$. The angle of friction θ_s is the maximum value for the brick staying in place; it provides a direct measurement of μ_s, the Coulomb coefficient of static friction.

This coefficient may be close to 0. This is the case, for instance, for two plates of teflon used to facilitate sliding a heavy piece of furniture on the floor. At cocktail hour, it's easy to see two ice cubes in contact, lubricated by a layer of water, slip and slide without encountering any resistance. There's nothing to keep this coefficent from exceeding 1, as when the super ball mentioned earlier comes into contact with a firm surface (or in the case of *Formula One* tires, which can reach 1.5 but wear out quickly). The raised coefficient of friction implies that the super ball prefers rolling to sliding.

Guillaume Amontons renewed these observations in 1699. In 1785, Charles-Augustin de Coulomb, who is better known for his work on electrostatics, did the same. As a military engineer, Coulomb started by studying the slopes of embankments of granular matter to determine the laws governing static and dynamic friction. Beyond a maximum angle, avalanches occur, which take the pile back down a few degrees, into a stable configuration. This stable angle is called the *angle of repose*.

*Coulomb's friction** depends solely on the interface between the materials sliding over each other, and is *independent* of their apparent contact area and normal applied force (figure 5.3). It might seem more likely that the force of friction would be larger when the contact surface is also larger. However, it wasn't until the mid-twentieth century that scientists understood that the contact area does not affect the coefficient of friction. Now, they recognized that the irregularities of contacts, which are always present on a microscopic level, and their plastic deformations, are the real cause of friction. Likewise, inasmuch as the force of friction can be seen to arise from entanglement between surface asperities, it would seem reasonable to think that the number of asperities increases with area, which contributes to resistance to sliding. However, this perspective does not consider the fact that the asperities become deformed under the effect of normal force—as we saw apropos of collisions between two grains—and that the actual contact area between asperities differs from the apparent area between objects (figure 5.5). When weight is increased, the actual contact range of all these points increases, too, which one can picture as two maps of mountainous terrain in relief, placed face to face. The contact area grows in proportion to weight (or normal applied force). Contact asperities form boundaries partially "welded" by van der Waals

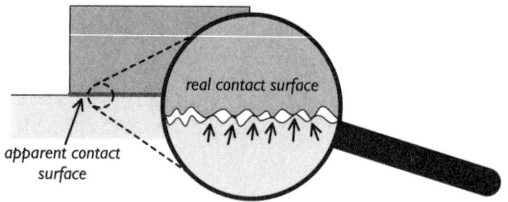

Figure 5.5
Two "flat" surfaces in contact. At higher magnification, the deformations of surface roughness are responsible for the phenomenon of friction.

forces. In order to undo them, a tangential force must be applied, which grows as a function of the tensile strength of boundaries and their number. In other words, frictional force increases in proportion to the actual area, which increases in proportion to normal force, accounting for Coulomb's law of solid friction.

ROLLING AND PIVOTING

The invention of the wheel certainly represents one of humanity's greatest achievements. In contrast to sliding material, which involves movement against the interface between two grains, *rolling* enables rounded grains to avoid shearing asperities. Rolling predominates in the deformation of most granular materials. Despite the friction, granular matter can flow like a liquid. That's why walking on a gravelly slope can prove to be as hazardous as sliding on ice!

Yet again, the game of billiards will allow us to understand more fully the different relative movements of a sphere. Varying levels of force, points of incidence, and aim transmit different kinds of motion to the ball. The kinetic energy of *rotation* is communicated through a leather tip attached to the end of the cue. The coefficient of restitution between two balls is higher than 0.9; between a ball and a side wall, it is about 0.6.

The simplest kind of motion is rolling without sliding. That said, a ball can slide and roll at the same time. In this case, it can spin in the direction opposite to its forward motion (*backspin*) or spin in the same direction, at a faster rate than its advance (*frontspin*). Pivoting involves the ball's rotation around its vertical axis, the way a top turns. After the player has performed the stroke, the ball's kinetic energy diminishes at each collision with other balls and sidebands, as well as from friction resulting from the table's surface.

When a weak, horizontal impulsion is applied to a billiard ball, it does not slide, but starts to roll. During this motion, the asperities at the point of contact between the ball and the table's lining are constantly being broken and reforming. This coefficient is markedly lower than the coefficient of sliding friction we observed earlier. For billiard balls, it is on the order of 0.01, twenty times smaller than the sliding friction between the ball and the table. This low dissipation value accounts for why rolling is the principal mode of relative motion between grains.

A final mode of relative motion for a ball that dissipates little energy is *pivoting*. In contrast to rolling, pivoting involves sliding between the ball and the table's lining, but it is a matter of rotation rather than forward motion. Since the contact area has a limited extension, a large number of revolutions is required before the rotational energy is spent. The same holds for a well-made top. In conjunction with rolling, pivoting is a relative mode of movement that is quite common in particles of granular matter in the course of deformation.

Numerical simulations of soil deformation on an inclined plane have shown that for 90% of contacts, the grains roll; only 10% slide. In the latter group, a large quantity consists of contacts where the normal force— and therefore the tangential force of friction—is weak. These observations suggest that granular matter deforms in such a way as to minimize the rate of energy dissipation by friction.

ELASTICITY OF A PACKING

As we have seen, small elastic and plastic deformations between grains govern force at the point of contact. We will now examine a pile with a large number of contacts. Just as we examined the relation between grain compression and contact flattening, we now turn to the relationship between the force applied to an assembly of spheres and the overall, elastic deformation that results.

For a clearer idea, consider a packing of beads in a cylindrical vessel, with a force applied to the top beads by means of a piston. The packing's mechanical response depends on the initial arrangement of grains. It may differ from one sample to another, according to how the vessel has been filled. In schematic terms, we can observe three regimes, depending on the intensity of the force applied:

- When force is weak, a major deformation of the packing occurs, resulting from local displacements of grains; this takes place especially at the interface between the piston and the grains, because the upper surface of the pile is not level, even after pressure has been applied repeatedly. In this regime, the law of deformation at the points of contact has little significance. Applying force has the effect of increasing the packing fraction in an irreversible manner. This is known as *granular plasticity*, which we will discuss in greater detail in chapter 8 with regard to wet grains.

- When force is more pronounced—that is, under a regime of intermediate force—deformation on a local level is mainly governed by Hertzian contact between grains. New contacts between grains progressively come into play when force is increased, and grains draw together. It should be emphasized that it is the number of mechanical contacts transmitting forces that increases, not the number of geometric contacts. The densification of the stress network in an experiment conducted by Pierre Dantu, which we will describe in chapter 7 (figure 7.1), exemplifies this state of affairs. Under this regime, the packing exhibits greater and greater resistance as mechanical stress is exerted: applied force increases more rapidly when there is deformation than in the case of a single Hertz contact. This fact is explained by new contacts made when compression is increased.

- At a still higher force, the number of contacts conveying pressure hardly varies at all. On the global level, one expects a confirmation of Hertz's law. More often, however, the problem follows a different course: the response, at the level of individual grains as much as that of the packing, is not elastic; here, the domain of plasticity—or even the point of grain rupture—has been achieved. In either case, pressure irreversibly modifies the grains themselves, crushing or breaking them.

Compression of an Array of Spheres

The relation between force and displacement (as discussed in a previous inset on Hooke's law) for a granular medium subjected to compression can be formulated as a relation between the axial stress $\sigma = F / A$, where A represents the area of the cylinder's cross section, and the axial deformation $\varepsilon = \delta / L$, where L stands for the cylinder's height. The pile's elastic behavior translates into the relationship $\sigma = K\varepsilon$; K denotes the packing's compressibility. The relation $F \sim \delta^m$ now indicates that compressibility increases with stress as $K \sim \sigma^{(m-1)/m}$. Thus, in contrast to most familiar materials and even in the case of small deformations, the compressibility of a granular medium depends on the confining pressure.

THE PROPAGATION OF SOUND

EARS FULL OF SAND

Desert scorpions have sensory organs located below their pincers that enable them to detect prey at a distance. Blind moles in the same habitat have this ability, too, which they perform by plunging their heads underground. The sand propagates the acoustic waves made by animals' movement. These elastic waves, as if a compressed coil spring were being released, occur because of very small periodic deformations of grains. Given that this environment is very disordered, one might wonder whether grains really propagate elastic signals. To understand the process, it is not enough for contacts to exist; sonic vibrations will open them if the grains are not sufficiently confined. Thus, the grains have to press against each other for the vibrations to propagate in the medium. By taking precise measurements at the entrance and exit of a vessel filled with grains, scientists have shown that vibrations travel down all possible paths, producing a highly chaotic signal known as *speckle*. Despite this source of fluctuations, it is possible to detect a *coherent* signal, the sound transmitted by the medium. Thus, even if disorder scatters sonic waves a bit, the assembly of grains behaves as if it were an elastic solid.

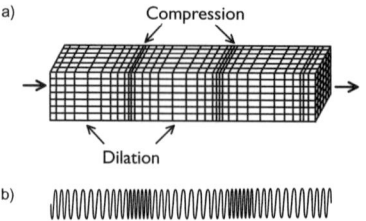

a) Compression

Dilation

b)

Figure 5.6

a) Propagation of a compression wave in a solid. b) Slight longitudinal deformations spread as in a spring that is periodically compressed at one end.

Sound

If the surface of a solid block is subjected to periodic compressions, they will travel inside the solid, just as a coil spring made to vibrate periodically at one end will communicate the periodic longitudinal vibration along its axis (figure 5.6). This sonic mode also exists in a liquid. Another mode involves the periodic shearing of the medium. Sound propagation in shear flows only occurs in solids; it plays a minor role in granular media. The linear relation between the stress associated with periodic shearing and the induced motion is expressed in terms of a shear elastic modulus, which plays the same role as the compression modulus K. Shear waves allow the solid character of a medium to be identified, but they are often more difficult to produce than compression waves.

This coherent signal in a granular medium results from the existence of the effective elastic modulus of compression, K, responsible for the propagation of waves. As is the case for all solid materials, the speed of sound is determined by the square root of the ratio between modulus and density: $\sqrt{K/\rho}$. Its speed grows with compression and diminishes if the density of the grains increases. K increases with confining pressure, which means that the speed of the sound increases, too. For this reason, the speed of sound in a granular material subjected to weak pressure (for instance, sand near the surface), and under the effect of its own weight, can be markedly lower than the speed of sound in grains, and even as low as the speed of sound in the air. However, since the pressure resulting from the soil's weight increases with depth, the speed of sound also increases. Thus, the waves become deviated (refracted), as they make their way through the medium, toward the surface. This effect is analogous to the mirage that occurs when light is deflected as a result of differences in the air's optical index stemming from higher temperatures near the ground. The sound waves in a soil are reflected downward when they reach the free surface, and thus they continue to spread along the surface, like ripples on the water. These slow, low-frequency waves can travel great distances—and come to the attention of desert scorpions!

THE ACOUSTICS OF GRANULAR CRYSTALS

A continuous chain of metal spheres will allow us to examine this wave effect. "Newton's cradle" consists of a series of spheres hanging individually on strings, such that they are barely touching (figure 5.7). The chain is brought into motion when one lifts, then drops, the ball at one end. The pressure of collision passes from one sphere to the next: the process involves a local wave whose speed depends on the initial force of impact. The speed of this shockwave tends toward zero when the amplitude of initial displacement is weak.

But if a fixed pressure is applied, an impulse is propagated that is comparable to exerting static pressure. A wave propagates in the medium. It's the kind of wave that scorpions sense, which permits them to locate prey by estimating direction and distance.

This is a simple example of a metamaterial. In chapter 2, we described another type when discussing opals and photonic crystals. Today, materials of this sort can be manufactured in one, two, or three dimensions, reproducing atomic models with larger elements in a structure grain by grain or layer by layer.

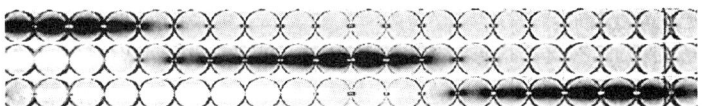

Figure 5.7
A chain of photoelastic grains viewed at three successive moments after impulsion from the upper left. The speed is practically constant. Light Hertzian deformation of grains is evidenced by the gray color on the grains.

It is possible to control the propagation of an acoustic wave in a three-dimensional *granular crystal*. Doing so involves modifying the pressure applied to grains while carefully combining grains of the same size and different mechanical properties, or introducing packing defects. These materials, constructed ad hoc, have remarkable acoustic properties. They can block the propagation of sound or, alternatively, deflect and focus a wave. To take just one example from this new field: acoustic lenses are being made that enable an acoustic wave to be focalized in a specific part of a human body, for noninvasive surgical procedures. Given that it is now possible to construct them grain by grain, these metamaterials promise untold applications in the future.

PLASTICITY OF PACKING

If a sufficient traction is applied to it, a copper wire will extend irreversibly. This is an instance of *plasticity*. The same holds for a granular material to which enough shearing is applied to prevent it from returning to its initial form. Plasticity on the level of contact between grains results from friction. On the scale of a packing, however, it is caused by the relative motions of grains.

A granular material confined by pressure P (figure 5.8) can be deformed in irreversible fashion by shearing, which resembles the friction of a pad sliding on a plane (also see figure 5.4a). The *shear stress* τ produces a *shear deformation* ε that increases over the course of time t. Shear deformation ε is the ratio between horizontal motion Δx and the thickness of sheared grains h. Experiments show that τ is proportional to pressure during shear. This defines the *macroscopic* coefficient of friction as $\mu = \tau / P$. As

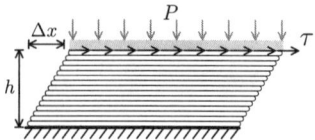

Figure 5.8
A sample of grains of h thickness under normal pressure P and sheared by tangential stress τ moves by Δx. The image is analogous to an ensemble of sheets of paper sliding on top of each other.

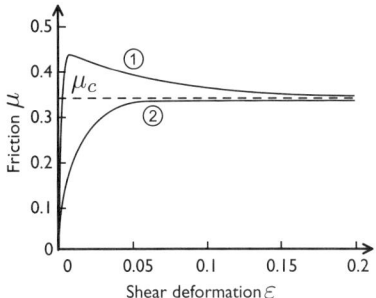

Figure 5.9
Evolution of friction coefficient μ as a function of shear deformation ε for a granular material that is initially dense (curve 1); and one that is initially loose (curve 2).

with the solid friction that occurs with a sliding pad, this coefficient is independent of P, but its value varies with shear deformation.

Figure 5.9 shows the value of μ as a function of shear deformation for two different initial states: a dense state (where the grains are organized in a compact assembly) and a looser state. In either case, the coefficient μ varies with the progressive evolution of the grains' configuration. At first it is small, but it increases during deformation, then tends toward a constant value μ after reaching a 20% deformation.

For the moment, let's consider only the value $μ_c$, which is reached after some shear deformation and is independent of the initial state. It is known as the coefficient of *internal* friction of the granular material. This property, intrinsic to the material, characterizes its shear strength, or, in other words, the value of a friction coefficient for a steady and slow flow. Despite the analogy between the friction coefficient $μ_s$ of individual grains and the friction between two layers of grains, the value of $μ_c$ is different from $μ_s$—a difference explained by the fact that not all the grains between two layers slide against each other; instead, a majority of them rolls. The relation $μ_c = \tan φ_c$ defines the *angle of internal friction* $φ_c$. For most sands, this angle lies at about 30 degrees, but it can reach 40 degrees in construction materials consisting of highly angular grains.

Whatever the initial configuration of grains may be, then, these rearrangements during shearing have the effect of bringing the material toward the same state, with the same value for $μ_c$. This is the *critical state*, which we discussed in chapter 4 when the packing fraction reaches a constant value, independent of the initial state. Thus, critical state is characterized by both a critical packing fraction and a critical coefficient of friction.

Two examples will illustrate the effect of a packing's mechanical strength when subjected to pressure.

- The first, concerning proportionality between the shear and normal stresses, is the difficulty of planting a parasol in the sand at the beach! Only near the surface can the sand be easily deformed. The normal stress acting on the parasol is proportional to a *lithostatic** pressure that increases with depth. Consequently, the shear strength also increases with depth and makes penetration difficult. This is one reason it's hard to get enough shade on the beach!

- A second example concerns the mechanical resistance of vacuum-packed coffee. External atmospheric pressure exerts a pressure on the package, and therefore on the grains. If you let air in through a small hole, you'll notice that the package's resistance to deformation progressively decreases. This simple observation illustrates the vast difference between a classical solid and a dry granular medium. In a granular medium, this network of contacts and the forces of grains that are not fused together can only form under the effect of confinement. The confining stress plays the same role for the packing that compression does between two grains. It can simply come from the weight of grains, or from forces acting at the boundaries of the assembly. For instance, when grains are poured into a cylindrical vessel, they are simultaneously confined by their own weight and by reaction forces at the vessel's walls. Increasing this stress can produce two effects: First, the grains will rearrange themselves. Second, all the forces between grains and the confining forces at the container's wall will increase in proportion to the stress applied; the grains become deformed without moving.

Let's return to figure 5.9. As we have seen, shear stress depends on the initial packing fraction. In the case of a loosely packed sample, the coefficient of friction, μ, increases with shear deformation and tends toward μ_c. In the case of the dense sample, μ rapidly increases toward a maximum value, often called "peak stress," before it declines toward the critical value μ_c. In contrast to μ_c, the peak stress does not represent an intrinsic property of granular matter; the higher the level of initial packing fraction, the larger it gets. This is why civil engineers try to obtain maximum soil density when building, in order to secure the best possible resistance to weak deformations. In contrast, engineers in the domain of powder technology are interested in the large deformations at work in particle flow; here, they pay attention to the critical resistance μ_c.

As for the critical state, we can define a peak coefficient friction μ_p from the ratio of the peak stress to the confining pressure, as well as a peak friction angle φ_p by setting $\mu_p = \tan \varphi_p$. This also represents the angle of stability for a slope made of the same granular medium (as the inset that follows explains).

Angle of Stability

Look at the side of a dense slope inclined at angle θ to the horizontal axis (figure 5.10). It can be divided into several parallel layers of thickness, h, with each layer capable of sliding on the one below it. In the upper layer, the weight per surface unit of grains is given by ρgh, where ρ represents the density of the material and g is the acceleration of gravity. This stress can be decomposed into components perpendicular and parallel to the layer. The component perpendicular to the surface is the confining stress, and its value equals $\rho gh \cos \theta$. The component parallel to the surface is the shear stress, $\rho h \sin \theta$. The ratio of the shear stress to the normal stress should be below the peak coefficient of friction $\mu_p = \tan \varphi_p$, which implies that $\theta < \varphi_p$. This means that a slope's angle of stability cannot exceed φ_p.

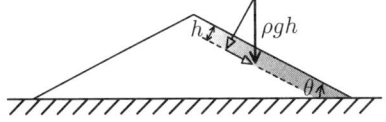

Figure 5.10

A granular heap and stresses exerted by a superficial layer on the surface underneath. ρgh represents the pressure exerted by the layer in gray, which can be decomposed into two components (normal and tangential, indicated by open arrows) on the sliding layer.

LOCALIZATION OF DEFORMATION

Until now, we have assumed that a material behaves in a homogeneous fashion at all levels when being sheared. Homogeneity means that if, for example, the volume changes as a consequence of shearing, the relative change will be the same throughout. This is not generally the case, however. In the event of a landslide or avalanche, it is possible for the sliding to occur only in a small layer (a sheet of snow, for instance). Then, the material is sheared in an inhomogeneous manner, and it can reach a critical state more rapidly in parts where shearing is stronger. It also leads to a difference of packing fraction at different parts of the material as a result of dilatancy. Under such conditions, the relationship between stresses and deformations measured for the assembly depends on inhomogeneities in the thickness of a given layer. We can observe this effect by applying horizontal pressure to a pile of sand with a bulldozer. If the pile is dense enough (with a packing fraction above its critical value), the sand will divide into two almost rigid "blocks" with one of them sliding on top of the other along the layer separating them. The deformation is said to be localized in a *shear band*. Figure 5.11a provides an example. Since dilatancy is proportional to shearing, the packing fraction in this band is lower. Accordingly, it is easy to observe the sheared zone by its packing fraction, which is lower than in the rest of the medium.

When deformation is localized in a band, it may be said—if we call to mind a moving pad sliding over a surface—that one layer of the granular matter is sliding over the rest of the material. But in contrast to the moving pad, which slides over a well-defined plane, the shear plane in a granular medium takes place within the material itself and displays an orientation that depends on the orientation of stresses imposed on the material. Shear bands slightly different in nature also occur on a large scale in the form of geological faults, which as a rule concentrate seismic motion (figure 5.11b). On the basis of such bands' orientation, geologists are able to determine the stresses that caused them. The geometry of shear bands tells us about the circumstances that created them, which can also vary over time. The history of the Earth is also a history of granular matter!

Figure 5.11
a) Shear bands (localized porous zones where material dilates) in an experiment of compression. The upper wall is displaced downward, while the lateral sides maintain horizontal pressure. b) The parallel bands of different shades of gray on this rock cut for a roadway are traces of localized deformations from sheared layers.

Having observed grains, having piled, packed, and stacked them, and having described the geometrical organization of model granular structures, we now have seen how their organization affects their packing fraction. We have also considered the small displacements and deformations of grains subjected to stresses. The following chapters will examine how the disorder in such organization governs the properties of packings.

6

MANIFESTATIONS OF DISORDER

Order is the pleasure of reason, but disorder the delight of imagination.
Paul Claudel

Clafoutis, a French dessert like flan, offers an example of a heterogeneous system where cherries are distributed in a disordered manner in a batter (a soft solid). Why not just serve a custard with a bowl of cherries on the side? The idea is that by mixing ingredients, new flavors and consistencies will result, yielding a whole that is tastier than the sum of its parts. This chapter examines disordered systems that constitute something more than a mere addition of elements. In so doing, we will identify emergent behaviors that are useful in applications of granular media.

Our starting point is a simple observation: granular matter is heterogeneous in more than one way. It is composed of wholes and voids, grains and pores. The latter may be filled by one fluid phase or more: air, water (in the soil), or hydrocarbons (in oilfields).

Heterogeneity also concerns contacts between grains, which play an essential role for understanding the electrical and mechanical properties of granular materials. We have seen how ambiguous such contacts can be, depending on what is being measured and the microscopic nature of the material itself. At any rate, it's not a simple matter of all or nothing. Electrical resistance between two grains can vary between zero and millions of ohms with just a slight deterioration of contact.

One encounters a third source of heterogeneity in mixtures of grains of different physical natures. We will turn to a situation of this kind in the context of the mathematical concept of *percolation*.*

This chapter will examine electrical effects in an assembly of conductive grains. Chapter 7 will then take a detailed look at mechanical problems in granular packings.

SCALES OF DISORDER

Heterogeneous matter may be studied on different scales, handily summed up as "3M": *Micro*, *Meso*, and *Macro*.

Micro refers to the scale of a single grain, or a tiny quantity of grains. If we know their geometry and physical properties, we can examine how deformation will occur if they are compressed, or if a contact between two grains is modified. In chapter 5, we showed how such disorder is constructed, and how the local properties of grains and contacts between them affect organization on a large scale.

Events on the *meso* scale—or, more precisely, on mesoscopic scales—represent a matter of particular interest in the following. A block of sandstone or a bag of marbles gives an idea of the scale in question. Something this size is much larger than a grain, but it's still small enough that its average properties (packing fraction, for instance) will not vary significantly in different parts of the ensemble.

On the *macro* scale, physical properties differ between regions in a given sample. This is the case for a soil with a composition that varies depending on depth, or for a construction.

An essential question is how, on the intermediate *meso* scales, a heterogeneous material can be represented by an equivalent homogeneous medium. To this end, let's examine the density of the medium, defined as the volume fraction of grains. Inside a sphere of radius R, it is measured from the central point, as represented in figure 6.1. The density has a constant value as long as R is smaller than the radius of the spheres. Then, it fluctuates inasmuch as R increases to incorporate more grains and more empty zones. Finally, when the radius is sufficiently large—typically, when it's three or four times larger than that of a single grain—the density inside the sphere reaches a constant average value with respect to full

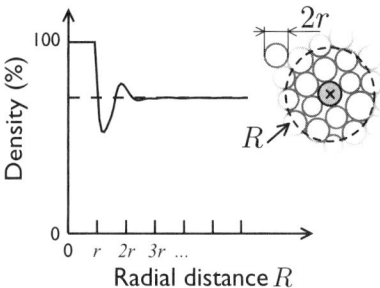

Figure 6.1

The average density of a pile of spherical grains of average radius r is evaluated in terms of a (larger) sphere of radius R centered on one of the grains. For a value of R lower than r, the material density of the central sphere is measured. The first maximum density for a distance equal to 2r corresponds to a sphere englobing only the centers of spheres in contact with the first sphere. When the value of R is high, the mean density is constant and lower than the density of spheres, because of voids.

and empty spaces. We speak of *representative elementary volume (REV)*,* the minimum volume at which the density or any average value of the property under consideration reaches a constant value. If this average value does not vary too much when larger volumes are considered, one is said to be dealing with a *small disorder*. In the opposite case, which we will discuss a bit later on, the expression *long-range disorder* is used and no REV exists.

SMALL DISORDER

The representative elementary volume defined earlier allows us to understand average properties, which we will use to describe behavior on a larger scale.

If the medium exhibits sufficiently low disorder that it can be replaced with an equivalent periodic structure, there are methods for shifting from calculating properties at the level of the grain and its immediate surroundings to calculating those of a large packing of grains. The principle underlying what is known as *homogenization** is as follows: one defines an elementary cell (corresponding, for instance, to the elementary lattice of a crystal), for which the desired parameter is determined precisely. Then, this result is repeated periodically for the ensemble of grains in the medium.

Other approximative methods exist for studying the average properties of a weakly disordered material, based on the work of the Italian physicist Ottaviano Mossotti in 1850, which were taken up by his younger German contemporary, Rudolf Clausius. These two scientists set out to calculate the average dielectric constant of granular material dispersed in a matrix (like the cherries in a clafoutis) of material with a different dielectric constant. The methods they pioneered have undergone major developments. We can understand the advances by way of an analogy: picture a section of a pointillist painting by Seurat displaying a mauve color due to an array of little red and blue dots at regular intervals. If we wanted to replace this composite color with a single, uniform hue, we would put uniform mauve dots in the place of the red and blue spots, adjusting the composition until, when viewed from far enough away, it's impossible to tell the difference between the homogeneous color and the composite one. This method is called *self-consistent*: a color has been chosen that does not jar with the percentage of blue and red spots the painter used. In physics, replacing diverse elements with an average element is known as a *mean field* approach.

EXTREMELY LARGE-SCALE DISORDER

The other limit opposite to *macro* scales is when a heterogeneous material is sufficiently extended that the average properties (on the macro scale) undergo variations from one zone to another. One example is sandstone, a rock composed of bonded grains whose density increases continuously with depth in response to so-called *lithostatic* pressure exerted by upper layers. It proves quite difficult—if not impossible—in such a case to find average laws that hold for the ensemble. If larger and larger volumes are necessary before reaching a constant mean value, one cannot define a homogenization volume. The discussion to follow does not consider such instances of *extremely large-scale disorder*, which vary from case to case.

LARGE-SCALE DISORDER

Between these limits of disorder, there are materials for which the size of a representative elementary volume cannot be defined. In such a case, one speaks of large-scale *disorder*. The flat colloidal aggregate represented

in figure 2.9 provides an illustration. The density ρ of this aggregate—
that is, the number of individual grains per unit of surface—can be determined by tracing increasingly large circles with the radius R from the
center of a reference grain, as in figure 6.1. But unlike the case of small-
scale disorder, its density does not tend to a constant value as R becomes
sufficiently large. In the example of the colloid in figure 2.9, mass M as
a function of R is proportional to R^D, where $D = 1.75$ is the aggregate's
fractal dimension. Its density, $\rho = M(R)/R^2$, is proportional to R^{D-2}. It
decreases as R increases: inasmuch as the radius is smaller than the size of
the colloidal aggregate, no average can account for heterogeneities present on a larger scale; an aggregate of extremely large dimensions would
have a density of almost zero. *Les Fractals*, a cartoon by Ian Stewart, tells
the story of a clever and dishonest cheesemonger who makes swiss cheese
with fractally distributed holes: measuring them by volume, and not by
weight, he sells more and more empty space in proportion to size! The
fractal geometry of percolation, which will provide a guiding thread for
the rest of the chapter, represents another case of large-scale disorder.

PERCOLATION, OR ARIADNE'S THREAD

A MODEL EXPERIMENT

The effect of large-scale disorder is easily illustrated by means of a table-
top experiment. It was first performed by an isolated group of teachers
who did not have the financial means to conduct experimental research.
They knew about percolation, a very general mathematical concept relat-
ing to disordered systems. It had been discovered by the British mathematician John Hammersley when he was investigating the progressive
clogging of gas masks and subsequent loss of *permeability** (chapter 11
will return to this matter). For their granular medium, our researchers
chose the hard spheres of sugar used for decorating cakes—the kind that
often come with a silvery coating, but can also be bought without it.
They made a set of batches, containing variable percentages, p of conductive spheres. Since the silver coating is very thin, the geometry of
the two kinds of grains was practically the same, and the grains mixed
without any segregation effect. Then, the researchers put the mixtures
into a large cylinder one after the other and compressed them by means

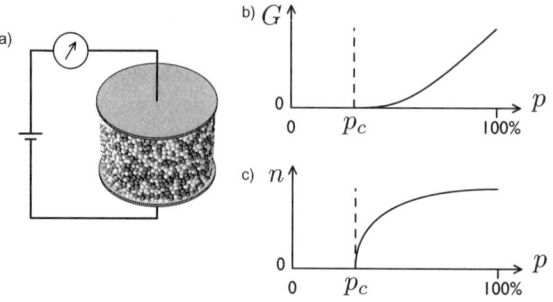

Figure 6.2

This is an experiment easily performed at home. First, a), a mixture of conductive grains in proportion p and insulating grains in proportion (1−p), which are geometrically identical, are compressed in a cylinder between two flat electrodes. Then b) the percentage p is made to grow from 0 to 1. The value of conductance G is non-zero only above the percolation threshold, p_c, and varies continuously up to the value contained when all the grains are conductive. Just above the threshold, the variation of G with the distance to the threshold $(p - p_c)$ does not depend on the detailed nature of the system, and in this sense, it reflects the "critical" nature of the threshold. Finally, c) the fraction P(p) of grains belonging to the conductive cluster (percolation cluster) varies much more rapidly than G because of many free strands belonging to the conducting cluster; its dead arms are not involved in conductance.

of a car jack in order to ensure good electrical contact between the conducting fraction of grains. Finally, they measured the mixtures' electrical conductance as a function of the percentage, p, of conductive spheres (figure 6.2a). Should all the spheres be conductive (p = 1), the current would pass freely from one end of the cylinder to the other. In contrast, if the percentage of conductive spheres was too low, the medium would act as an insulator. The team paid particular attention to what happened when *p* was near a *threshold* concentration—let's call it p_c —at which the system switches from an insulating regime to a conductive regime with only a few continuous pathways for transmitting the current. Figure 6.2b provides the variation in conductance G they observed.

A pretty humdrum experiment, you might think. In fact, it's not dull at all. The general properties of many problems that physicists study with attention to threshold values (like p_c) are independent of the physical system under examination. In this experiment, the value of p_c is close to 0.3; in other words, the system conducts electricity when more than 30% of the spheres have a silver lining. As this concentration increases, the number of conducting paths connecting the opposite electrodes increases rapidly. The most remarkable thing about the experiment is the fact that the law of the increase of conductance from zero as a function of distance $(p - p_c)$ to the percolation threshold is the same for different granular systems; in cases like this, physicists speak of *universality*.

In contrast, there would be no way to find the solution to the problem of percolation if one used a model based on averages. To calculate the conductance of an ensemble, knowing the nature of electrical contact between two or several grains is not enough. When it lies by the threshold, the geometry of conduction paths is a fractal object. In order to characterize the conductive medium, one should look at the full packing of grains: this is beyond homogenization!

OTHER INSTANCES OF PERCOLATION

The qualitative interpretation of this model experiment is simple. The electrical current travels on a continuous and tortuous path traced between conductive grains one by one (see figure 6.3)—like an inveterate cross-country skier who, at the end of the winter season, tries to steer along what remains of the snow so he won't have to stop at a patch of grass. But the effect of this phase disorder can also be formulated in rigorous scientific fashion.

Percolation accounts for various situations of this type.

- In tire manufacture, grains of carbon are included in the matrix of rubber before curing; this carbon contributes to the tires' mechanical properties.

- A jelly "takes" if the concentration of setting agents in the fruits or sugar employed is sufficient. The *threshold* value that makes the combination work (the gelling threshold) is reached when continuous chains of molecules appear that ensure an elastic behavior for the gel. Increasing the modulus characterizing the gel's elasticity is *analogous* to increasing the electric conduction in the mixture of conducting and insulating spheres by adding more conducting grains. Our reference

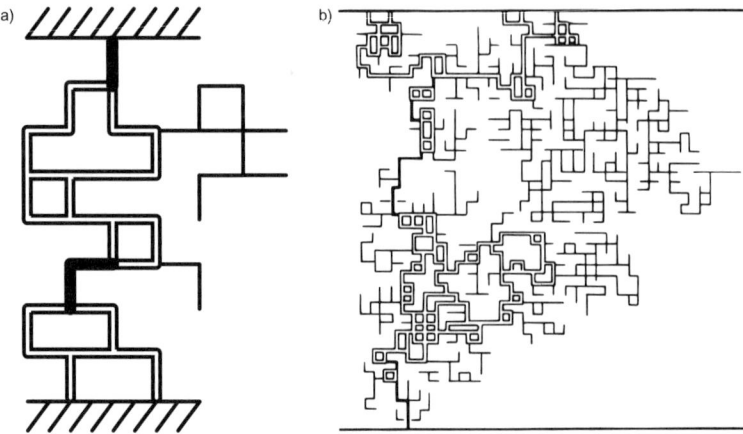

Figure 6.3

a) A random percentage of conductive bonds, p, precisely equal to percolation thresh-old p_c, is placed on a square network (in this case, $p_c = 1/2$). b) A similar image at large scale given by a computer simulation shows the form of an "infinite" cluster. The con-nections where the current's "traffic" is meant to pass—in bold—play a decisive role for conductivity, G (see figure 6.2b). Connections represented by double lines are other connections of the cluster crossed by electrical current. Connections in thin lines are "dead ends" where no current circulates.

to *analogy* means that there are rigorous laws of correspondence that allow us to shift from one problem to another, provided that each situation concerns the vicinity of a threshold.

In everyday life, "percolation" evokes percolators used to make coffee, the passage of a fluid (water) through a porous medium (grounds) from which the fluid extracts coffee. The term keeps the same meaning in the realms of geochemistry and chemical engineering. In this chapter, the mathematical notion of percolation also deals with a transfer within a random structure. However, a major difference in making coffee is that attention is paid to a problem around a *threshold*, a particular value of a parameter governing the structure's connectivity, like the critical concentration p_c in our conductive spheres. The proportion of electrically connected grains that can be made to vary in the spectrum between 0 and 1 is the *order parameter* of percolation.

PERCOLATION TRANSITION: A *CRITICAL PHENOMENON**

These two parameters (threshold and order parameter) are typical of changes from one state to another, like that of magnetism for an iron bar around a Curie threshold temperature: temperature represents the *control parameter*; a *threshold temperature* exists, below which magnetism appears. The region of higher temperatures, where magnetism vanishes, corresponds to that of low concentrations of conductive grains (such that no continuous path exists between them). Above the percolation threshold, a *long-range order** appears that takes the form of a continuous ensemble, the so-called *percolation cluster*, between grains in contact with each other and connecting both electrodes. The equivalent of magnetic ordering, for percolation, is its connectivity.

The geometrical description of the percolation cluster has been the object of many numerical experiments conducted with periodic lattices from which a given percentage of elements has been removed at random. These elements can be the vertices of a lattice, in which case we speak of *site percolation*. Alternatively, they may be bonds linking two sites; here, *bond percolation* is at issue. (Note that the mixture of conductive and insulating spheres represents a problem of site percolation.) Figure 6.3 represents the bond percolation at threshold for a square lattice. This cluster presents a tortuous structure that proves self-similar at any scale of observation; in other words, it's a fractal. Its *fractal dimension* is $D = 1.89$ in two dimensions, and $D = 2.52$ in three dimensions. In either case, it is less than the spatial dimension the cluster occupies. Fractal structure corresponds to large-scale disorder that cannot be described through homogenization. In contrast, the properties displayed near the threshold (the critical concentration or temperature), which depend on this geometry and not on local details, have general characteristics that define their Universality.

PERCOLATION THRESHOLD

Threshold values determined on different regular lattices exhibit great diversity: the value of the percolation threshold is no more universal than is the Curie temperature for magnetism in different materials; what is universal is how the order fits into the conducting (or magnetic) regime.

An estimate of the conduction threshold is made in terms of the volume occupied by the conducting phase in a mixture. For a surface problem, it lies at about 45%, and for a volume problem at about 15%. A much lower value of 8% is obtained in the case of carbon black, used in the isolating matrix for car tires. The formation of aggregates of particles accounts for the difference. In the latter case, porous grains made up of agglomerates of carbon microspheres, not the microspheres themselves, determine the geometry of the conductive cluster.

But more than the packing fraction, the quality of contacts or the percentage of conducting contacts between neigboring spheres—that is, the *coordination number* z (corresponding to the number of nearby bonds around one site)—determines the threshold value on a large scale. Thus, when bond percolation is at issue, one finds that it takes, approximately, the presence of two conductive bonds per site to ensure the lattice's connectivity on a plane. A lower threshold value found in three dimensions suggests that it is easier to establish connections in volume, that is, when the constraints of plane geometry do not hold.

PERCOLATION: GENERAL REMARKS

The notion of percolation permits us to shed light on the problems presented by conducting lattices with no threshold, by only using conducting bonds that present a very large distribution of resistances. How can one estimate conductance in cases like this? The answer is achieved, first, by suppressing all resistances, then putting them back where they were step by step—starting with small resistance and progressively adding larger resistance bonds. At first, the medium will be insulating because there aren't enough bonds to ensure continuity. At a certain point, however, adding one resistance will yield a connection. This value represents a value of reference for overall conductance and defines a pseudo-threshold similar to percolation. Beyond this point, adding higher resistances won't significantly change the value of the conductance of the full lattice.

This process may be conceived as analogous to rush-hour traffic. When estimating travel time, the part of the trip on express lanes doesn't count so much. Instead, stop-and-go traffic determines when you get home. It's the worst—but necessary—part that matters.

LARGE-SCALE DISORDER IN ELECTRICAL CONTACTS

AN ENSEMBLE OF CONDUCTIVE GRAINS

The way an ensemble of grains reacts to mechanical stress helps us to understand the properties of conduction in a pile of conductive grains. Put an ensemble of spheres that are all conductive (such as silver-coated spheres of glass) in a cylindrical container and subject it to pressure along the vessel's axis. Its packing fraction will eventually stabilize after a few pressure cycles. An initial decrease of resistance comes from suppression of insulating layers naturally present around metallic grains and from the flattening of rough contacts between them. This occurs in addition to the pile's reorganization, which creates new electrical contacts.

THE BRANLY EFFECT

One of the most spectacular manifestations of the heterogeneity of contacts occurs in a phenomenon described by Edouard Branly in a report to the Academy of Sciences entitled *Variations of Conductivity under Electrical Influences* (dated November 24, 1890). Branly packed metal filings in an insulating tube, between two metallic electrodes that exerted a slight pressure on the packing. At first, the resistance measured between electrodes was very high because of poor contacts between grains: under such conditions, it can reach values of a million ohms. However, if the packing is subjected to an electromagnetic field—for instance, by means of a spark from a lighter at a distance from the tube—the resistance falls to just thousands, or hundreds, of ohms. This state of strong conduction persists once the spark is gone. However, all it takes is a gentle shock to the tube for the quasi-insulating state to come back. (This experiment can also be performed in one dimension, with a chain of steel spheres of the same diameter.)

In this way, Branly highlighted action that occurs at a distance, without a material link. A few years earlier, Maxwell's *Treatise on Electricity and Magnetism* (1873) described the nature of electromagnetic waves. Fifteen years later, Hertz identified these waves experimentally by means of his resonator (a continuous metal ring between the extremities of which scintillation appears when the spark of a Ruhmkorff coil is applied). But despite this pioneering work, it took Branly's tube and filings—or *radioconductor*, as he called it—to "reveal" the waves and make electro-magnetic radiation available for applications. In its day, the radioconduc-tor played a key role in the development of wireless telegraphy. In 1898, Eugène Ducretet used it to make the first transmission between the newly constructed Eiffel Tower and the Pantheon in Paris; a year later, Gug-lielmo Marconi did the same across the English Channel.

The tube filled with metal filings is now known as a *coherer*. The English physicist Oliver Lodge gave it this name because conduction is accompanied by the formation—thanks to adherence between grains—of tenuous and winding chains extending from one electrode to the other, which call a percolation path to mind.

Numerous studies of the physical phenomena at play in the myste-rious Branly effect have been conducted for more than a century now. It has been conjectured, for instance, that electrostatic interactions occurring between grains provoke their displacement, encouraging the formation of chains whose elements remain in contact once the elec-tromagnetic field disappears, through molecular attraction. It has also been argued that the potential difference between neighboring grains prompts the insulating layer to break down and microsparks to appear, which fuse the grains together. Another theory proposes that the con-tact surfaces are so small that the current density running through them, which is huge, provokes local heat that is great enough for metal to melt.

A team of researchers from Lyon has recently advanced a likely expla-nation. First, they showed that having a tortuous geometry that the current exploits in order to pass through the granular medium in the con-ductive phase isn't a necessary ingredient for understanding the Branly effect. The same effect can be reproduced with a linear chain of steel spheres in contact; hereby, the pressure applied to grains follows Hertz's law. With this linear geometry, cycling between conduction and insula-tion is obtained, as in a three-dimensional granular medium. In turn, the team demonstrated the importance of a thermoelectric mechanism: under conditions of poor conduction, a very weak current passes through microcontacts between grains. This process creates a significant release of heat at a local level, which causes the grains to dilate and "pierce" their insulating surface, which provides the basis for weak resistance in the second phase. As it turns out, the effects of electrical and thermal conductivity, which are both due to so-called conduction electrons in a metal, are proportional to each other, independent of the nature of materials and the nature of contacts. Here we have a reasonable and, all in all, simple explanation for this mysterious effect.

A situation somewhat similar to the one in the tube with filings is found in zinc oxide (ZnO) varistors. The latter are electronic components used to protect electrical devices against surges; their resistance varies with the potential difference at their terminals. Such ceramics are obtained by sintering conductive grains of ZnO that have been mixed with minimal quantities of different metal oxides (see chapter 9 for a discussion of *sintering**). These grains are separated from one another by an oxide barrier that only conducts the current when there is sufficient tension between two grains. The winding path of the current then creates a short-circuit effect (used to ground lightning rods, for instance).

PERCOLATION IN MECHANICS

Just like electrical properties, the mechanical properties of a packing of elastic and conductive grains depend on the disorder of contacts between elements and their continuity. In particular, it is possible to account for the mechanical properties of an actual pile of grains presenting a broad distribution of contacts by removing a certain number of contacts at random (as we did previously, in the experiment modeling percolation, by replacing a percentage of conductive grains with insulating ones).

The question arises whether the electrical and mechanical properties can be related to each other. The answer is no. Mechanical stresses modify the positions of grains and therefore the geometry of contacts. The difference may be illustrated by means of a model. Replace the bond between two grains with a bar, and a link between three grains with two bars attached at the level of the central grain and free to pivot (figure 6.4). This yields a triangular lattice of conductive bars. Then, at random, remove a certain percentage $1-p$. The electrical conductance of the lattice is finite, provided that the percentage of conductive bars is higher than the threshold p_c (on a so-called scalar percolation model). In contrast, the mechanical lattice can be deformed as a result of what is known as a

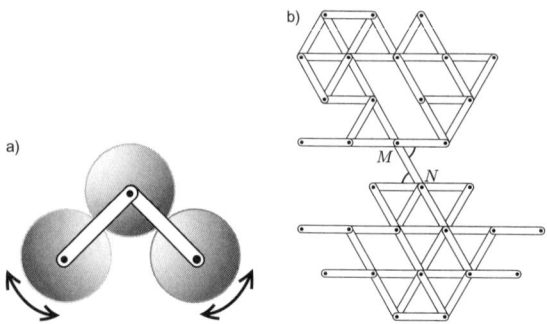

Figure 6.4

a) An ensemble of two articulated bars replacing an ensemble of three spheres in contact and free to roll. b) The mechanism simulates an incomplete granular medium. Locally, the upper and lower parts are rigid separately, because of triangulation. In contrast, the ensemble—while solid—is not rigid, because of the connection between *M* and *N*, which is free to rotate.

mechanism at this same critical bond number. An adequate, supplementary number of bonds must be added for the system to become rigid—which yields a higher threshold value for its elastic behavior. That would amount to a vectorial problem (with force vectors). The two problems are different. Looking at the Eiffel Tower and other nineteenth-century metal constructions, it's clear that these structures require steel trusses to stand firm. The matter was significantly important for even the great Maxwell to write about it!

However, vector percolation can account for the elasticity of so-called lattices comprising bonded grains—"consolidated" lattices—in which the application of force does not change the geometry of contacts, yet permits elastic deformations. They will be discussed in chapter 10.

Vector percolation is also found in other systems. Thus, a *gel* consists of a lattice of chains of polymers that are partially attached to each other, like a worn-out fishing net. If the mesh is adequate, the net remains continuous: it's above a percolation threshold. Moreover, it seems that mechanical properties around a threshold of gelation-percolation correspond to the electrical analog (unlike the example in the previous figure). This may be explained by the fact that systems in solution present effects related to osmostic pressure, which make the vectorial character of their mechanics disappear. As such, scalar percolation accounts for the transition from a viscous sol toward an elastic gel, as occurs when a fruit jelly "sets." To avoid wandering too far afield of our beloved grains, we won't go into further detail on this point.

We have arrived, at the end of our presentation of long-range disorder. Percolation, like our discussion of fractal geometries, has shown that it is pointless, in many situations, to try to understand the behavior of a disordered system on the basis of simply describing what occurs on a small scale. The validity of these models of long-range disorder stands beyond doubt. Instead of trying to find a rigorous application for them, we will see that they provide qualitative tools for obtaining prototypes of systems with many scales of disorder.

7

FORCE CHAINS

I can never satisfy myself until I can make a mechanical model of a thing. If I can make a mechanical model, I can understand it. As long as I cannot make a mechanical model all the way through, I cannot understand.

Baron William Thomson Kelvin

In a well-stacked pile of wood, the cylindrical logs in parallel seem to be touching. At the same time, however, some of them can be extracted from the middle of the pile because they're not blocked. This is a concrete example of the "arching effect," which manifests the unevenness of stresses at the level of contacts. This chapter will explore, in detail, how the forces at work in a packing are distributed.

VISIBLE FORCES

EXPERIMENTUM CRUCIS

The electrical conduction in a packed assembly of conductive grains has shown that this pile is something quite different from the simple sum of grains composing it. Now we will see that even knowing what happens on the level of a contact between two grains does not permit us to deduce the behavior of a packing of grains under compression. The most spectacular demonstration of this unevenness of compression forces was given

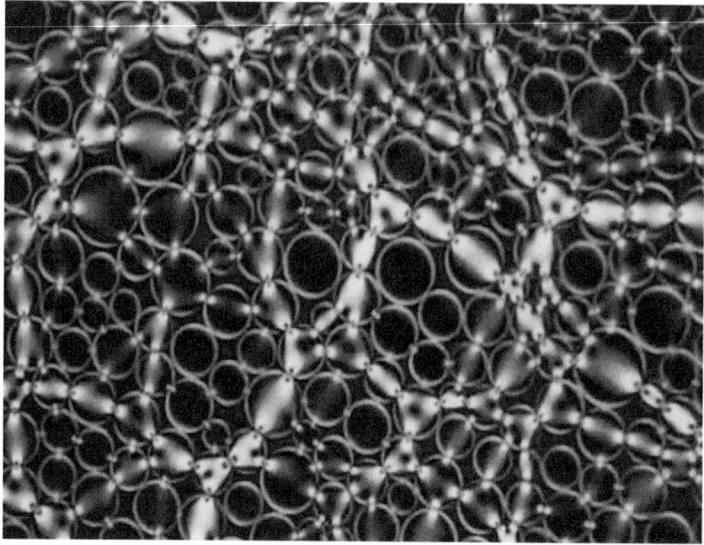

Figure 7.1

In a rectangular container, the transparent photoelastic cylinders here seen along their axis are subjected to a vertical stress. The irregularly distributed white lines visualize the paths of strong stresses. The dark looking cylinders are hardly compressed at all.

sixty years ago by Pierre Dantu, a civil engineer for roads and bridges. It involved applying pressure to piles of cylinders with parallel axes (this kind of model is frequently used in soil mechanics to simulate packings in two dimensions). Dantu chose plexiglass for the cylinders because of its properties of *photoelasticity**—in other words, the optical index of each cylinder inside a material is a function of the magnitude of the stress at that position.

The photo in figure 7.1 shows that, even if the packing appears to be globally homogeneous, the distribution of stresses in the grains isn't homogeneous at all. One can see an unbroken continuous filamentary network of luminous cylinders in contact, a so-called network of *force chains*.* This network grows denser as the pressure exerted on the packing increases, constituting the rigid skeleton that ensures the overall mechanical strength of the packing. We see that this skeleton surrounds regions where the grains experience little stress, or are even free of stress.

NUMERICAL EXPERIMENTS

Computer simulation is another means for investigating a granular material at a microscopic level (that of grains). Recent developments, especially those pioneered at the University of Montpellier, allow us to examine in detail the heterogeneity of granular media subjected to forces. In order to analyze the distribution of forces in a granular medium, the particles are confined by rigid walls. In a first step, identical pressure is applied on all walls. Under these conditions, the *shear stress*, due to tangential forces on opposite walls, is zero. In the next step, pressure is progressively applied between two opposite sides, and researchers observe how this stress is transmitted across the granular sample. The normal forces between the grains can be represented by segments with a thickness proportional to the magnitude of normal force, as in figure 7.2. As in experiments with photoelastic grains, one notices a highly inhomogeneous distribution of forces: many grains are subject only to very weak forces.

Figure 7.2 shows two examples of force maps: one for an assembly of discs, and the other for an assembly of pentagons. The segments connecting the centers of the grains reveal different contact networks. But in both cases, the number of contact forces lower than average force is independent of the magnitude of the force. It's easy to find contacts where the

a)   b)

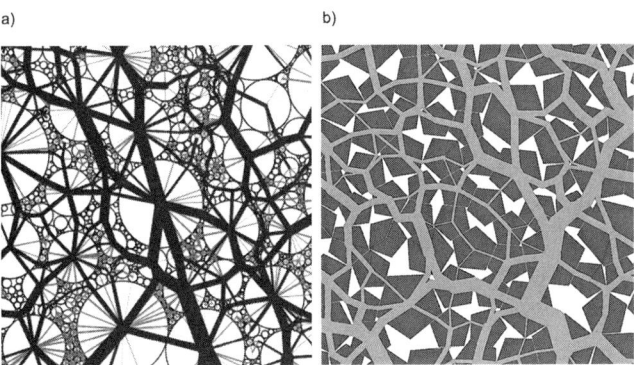

Figure 7.2
Force chains in assemblies of grains simulated numerically. The thickness of segments connecting the centers of grains are proportional to normal forces between grains. a) The grains are discs whose diameters vary by a factor of 100 between the largest and the smallest. b) The grains are polygons.

force transmitted is 10, 100, or even 1,000 times weaker than the average force. Such simulations reveal that near 60% of the forces are below the average force; these weak forces are not necessarily visible in photelastic records.

However, the distribution of forces above the average value follows a law of exponential decay. Exponential decay means that, in contrast to the distribution of molecule speeds in a gas, many forces exceed two or three times the average force.

WEAK VS. STRONG FORCES

The contact network in a granular medium, then, is composed of two distinct subnetworks with different distributions: a "strong" one, bearing forces higher than the average force, and a "weak" one, bearing forces below it. Simulations have shown that the two subnetworks make different contributions to the overall behavior of the packing. For instance, in the process of a slow deformation, the majority of grains rolls over neighboring grains without sliding. In general, there is only a tiny fraction of contacts where grains slide and dissipate energy; these contacts are in the weak subnetwork where normal forces between grains are sufficiently weak for the sliding to cost as little energy loss by friction as possible.

Knowing the positions of grains and the forces to which they are subjected by their neighboring grains, it is possible to calculate the stresses along all directions in space. The contact forces can be counted depending on whether they are weak or strong; as such, their contributions to stresses can be evaluated. The total stress inside the packing is generally a combination of an isotropic pressure (like the hydrostatic pressure in a liquid) and a shear stress. At this level, a surprising property of forces in a granular medium is shown by computer simulations: weak forces alone transmit almost no shear stress; In other words, these forces exert a simple pressure, as in a liquid, which represents a little more than a quarter of average pressure inside the packing. The strong forces, in contrast, carry all the shear stress of the packing and three quarters of the average pressure. Thus, this strong network acts as if it were a solid, in the sense that it bears all the shear stress and thereby enables the material to support different values of pressure in different directions.

In schematic terms, the distinction may be understood as follows: continuous chains of grains, almost aligned, exist in the granular medium. If they lie along the direction of the principal stress applied to the sample, such chains are able to transmit significant forces. These grains constitute the privileged links for the transmission of stresses. If they exhibit a weak misalignment with the direction of the principal stress (for instance, if a grain occupies a slightly staggered position), lateral support by a weak force will be necessary to compensate for this defect, and thus prevent the strong force chain from collapse. Accordingly, the weak forces that correct geometrical imperfections may have different directions and amplitudes, disconnected from the strong forces that brought them into being in the first place. This is why weak forces bear no shear stress, whereas strong forces reflect the overall shear stress to which the granular sample is subjected.

GRANULAR TEXTURE

The geometric disorder in an assembly of grains is dominated by steric hindrances: because of their volume, grains get in each other's way and cannot choose arbitrary positions in space. That said, in order to characterize granular texture, it's not enough simply to consider the positions of particles. Indeed, interactions between the grains are conveyed by the contact network, and the equilibrium of each grain is governed by the forces exerted on it by neighboring grains. Thus, to establish a connection between disorder resulting from position and the medium's mechanical behavior, the contact network and force network must be characterized simultaneously.

NUMBER OF CONTACTS PER GRAIN

In a network of contacts, the local environment of grains is described by the number of contacts, z, which each particle has with its near neighbors (i.e., in its surrounding layer of grains). This coordination number z varies from one grain to another, but it cannot exceed a maximum value z_{max}, because of steric hindrances arising from the volume occupied by each grain. Likewise, the equilibrium of forces exerted on a grain imposes a minimal number of contacts z_{min}. Since the equilibrium of a spherical grain subjected only to two contact forces proves unstable in three dimensions, z_{min} equals 3. By the same token, for spheres of the same size, z_{max} equals 12. Thus, in an assembly of spherical grains, the coordination number can vary between 3 and 12.

In a more general sense, the values z_{min} and z_{max} depend on the distribution of grain sizes, their shapes, and the nature of forces acting between them. The examples in figure 7.2a represent a system with a factor of 100 between the sizes of the smallest and largest grains. This range of sizes has a marked influence on the contact network: on one hand, the coordination number is greater; on the other, steric hindrances enable grains of small diameter to fill the pores left open between larger particles. Thus, it is possible for many small grains to be excluded from the contact network, that is, to have no contact with others; these free-floating grains, sometimes called "rattlers," are not involved in force transmission across the packing. Their proportion can exceed 50%, even if most of them are rather small particles that do not fill a significant volume. Observation also shows that the greatest forces are essentially borne by the largest grains.

For granular media composed of nonspherical grains (for instance, the polygonal particles shown in figure 7.2b), two particles may share more than one contact, or extended contacts, along their surfaces. The conditions of force balance are thus different from those of spherical grains. In some cases, it is possible to obtain packings that, while not very compact, possess a large number of contacts; this lends the material a greater resistance to shear, despite a low packing fraction. A similar observation holds for *cohesive* granular media (an example of which is found in figure 7.3). In contrast to noncohesive media, where a contact between two particles opens as soon as the particles are pulled, a cohesive contact between two particles resists applied tension. As such, the network involves two kinds of contacts: 1) contacts in compression and 2) contacts in tension. In the course of shearing, the latter type of contact occurs mainly in the direction of the material's extension, whereas contacts in compression appear mainly in the direction of the material's contraction.

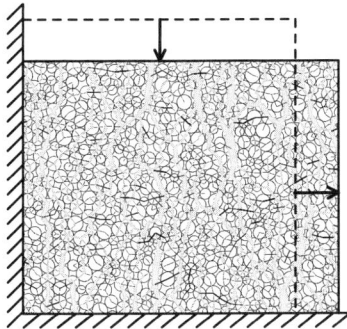

Figure 7.3
Computer simulation of force chains in an assembly of cohesive grains. Pressure is applied vertically to the sample, which pushes force chains (in gray) in a vertical direction, while also tending to separate them horizontally. To offer resistance, tensile forces between grains (black short lines) appear in the horizontal direction.

CONTACT DIRECTIONS

Granular disorder is also manifest in the highly variable contact orientations within the material. Contact orientation between two particles is perpendicular to the plane that is tangent to the two particles at their contact zone. On average, disorder tends to favor a uniform distribution of contact directions. Steric hindrances also favor isotropy in the contact network: in the layer of each grain's closest neighbors, jamming favors the existence of contacts in very different directions. In contrast, the weight of grains, along with shearing, lead to anisotropic networks in which a significant proportion of contacts occurs in specific directions. When the grains sediment, each one falls and comes into contact with another that has already been deposited. For noncohesive grains, this initial position is often unstable, and the particle will roll until it forms a contact with at least one other grain in the substrate. This process leads to fewer contacts in a vertical direction, and a majority of contacts lying in directions at 45 and 135 degrees to the vertical for an assembly of equal-sized particles.

When a packing is sheared, the grains move with respect to each other. In consequence, new contacts form in the direction where contraction is highest, and contacts open in the direction where extension is largest. In view of this anisotropic distribution, the average number of contacts between grains, z, only provides a partial description of the texture. It is also necessary to take the *anisotropy** of contact orientations into account. Recent studies conducted by means of numerical simulation have shown that the shear strength is proportional to anisotropy.

THE ARCHING EFFECT

You may have noticed that the flow rate of sand in an hourglass is practically constant until the upper part is almost empty. In contrast, the flow rate in a clepsydra—the water clock used in antiquity—diminishes continuously as the water level above the opening between chambers decreases. The response to the paradoxical behavior in granular media is the *arching effect*,* to which we now turn.

The accounts we have just provided of contact forces and networks have shown that a granular medium subjected to shearing assumes a structure in order to bear the stresses applied to it; in particular, the anisotropic character of the network and the formation of force chains reflect the arching effect. In an empirical manner, over time, medieval builders of cathedrals made arches that soared higher and higher in order to hold up the weight of vaults and provide more light to the structure's interior. The stability of such constructions does not depend on the mechanical qualities of the blocks of stone employed or adhesive materials, so much as the geometrical arrangement of the whole. Structural stability in Gothic architecture is ensured by an array of vaults that support each other, up to a keystone at the highest point (see figure 7.4a). In this way, the strain deriving from the framework and siding are efficiently transferred to columns and buttresses. Compare figure 7.4a with figure 7.4b, which shows a vertical force chain supported by weak forces applied almost horizontally on the grains through which it passes. In both cases, moreover, part of the structure is either empty (in the architectural vaults) or composed of bigger or smaller gaps, which contain grains not subject to any force. The number of these free-floating particles increases with polydispersity inasmuch as small grains can easily find a place in the large pores between big grains. This fact indicates that, both in a granular medium and in an architectural structure, forces are conveyed by markedly oriented elements (the strong force chains or columns in a

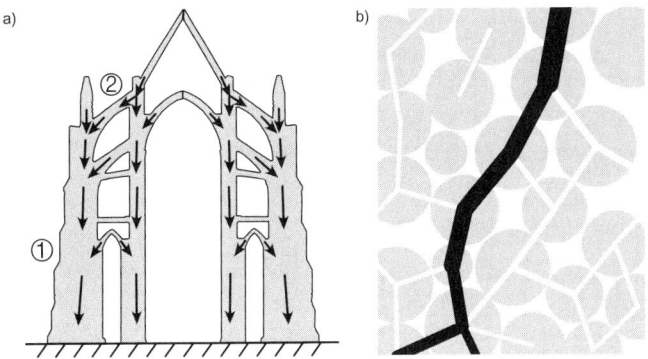

Figure 7.4

a) In addition to the weight of vaults, transverse forces are exerted that are counterbalanced by 1) buttresses and 2) flying arches. b) Forces in a granular medium are principally transmitted by highly compressed grain chains. However, a weak force network (represented in thin white lines) ensures transverse stability.

cathedral), which are supported by buttresses called counterforts (corresponding to the pressure exerted by the weak forces in a packing).

The stability of arches and their modes of rupture have interested engineers and builders for centuries. Figure 7.5 shows drawings by Augustin Danyzy for experiments conducted in 1732 on stone arches in Montpellier. Although it's relatively easy to make vertical walls with stone blocks, stacking them to make an arch involves mechanical difficulties. To do so, stones must be cut according to the function they will serve in the design. Even when the stones are fitted well, the arch may still collapse if the columns on which it rests are not firmly anchored on the ground with other blocks of the structure serving as support. When these rules are disregarded, either the stones will slide, or the contacts between them will break open to one side or the other. The condition for the arch's stability is that the force lines (represented on the first drawing by dotted lines) traverse all the stones composing it.

The foregoing analysis shows that, as with a pile of spherical grains, cubic blocks can roll and slide if the global geometry of the structure is poorly calculated. Clearly, the arching effect present in force chains is greater in angular blocks than in spheres. Figure 7.6 shows a packing of polygons in which the pressure exerted on each grain by its neighbors is represented in shades of gray. We can see grains under weak pressure

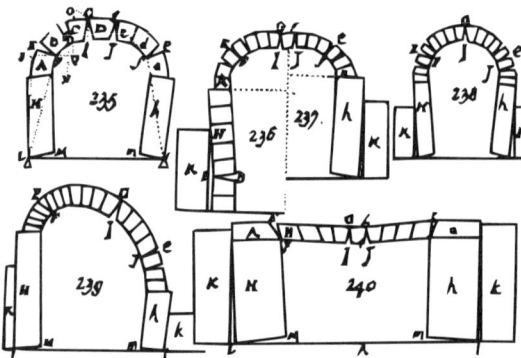

Figure 7.5
Sketches of experiments conducted by Augustin Danyzy in Montpellier to study the stability of arches.

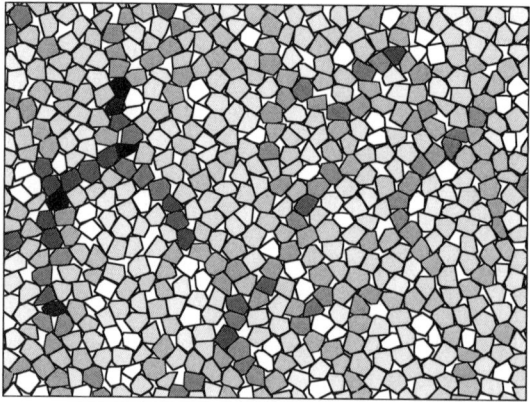

Figure 7.6
Pile of pentagons compressed by vertically applied force. This computer simulation indicates, by shades of gray, the distribution of stresses among individual grains.

surrounded by arches that are submitted to noticeably higher pressures. Recent studies have demonstrated that the stability of ballast, for instance, is related to force chains induced by the nonspherical shapes of the grains more than to the anisotropy of the contact network, as in the case of spherical assemblies discussed earlier. The effect resembles that of the "nuts" that rock climbers affix to cracks for their safety: the flat surfaces ensure very strong flat contacts between the rocks and the bolt.

If one pours a mass of glass beads into a container all at once, the resulting packing fraction is weak. Filling in this way leads to pockets and voids, sometimes of significant size. These voids are preserved by arches that may comprise several dozen grains. The arches are not very stable in the case of smooth spheres, which can easily slide and roll. A larger packing fraction is readily obtained by means of slight agitation or vibration— especially when it occurs in the process of filling. It's said that, in certain countries, merchants have been known to trick customers by plunging their measuring scoops into bags of grains quickly, in order to get away with selling less. Operating procedure makes a difference! There's no way to correct a poor initial packing of grains entirely. By applying pressure to a pile that isn't very dense, it's still possible to improve packing somewhat, because doing so provokes a certain amount of local motion. But if cycles of pressure or shaking are exerted, the packing fraction increases rapidly after just a few cycles—and a limit is reached, with each grain having come to occupy a stable position.

When the stability of contact between grains is high on account of the fact that rough surfaces restrict motion, or when the grains' shapes inhibit rotation (which would allow them to sink into a better arrangement), the arches formed at the time of filling prove quite stable: pressure on the pile will only reinforce the stability of arches by pressing the grains constituting them closer together. Accordingly, pressure is ineffective for increasing the grains' packing fraction (which, moreover, is weaker than in the case of smooth spheres). Unintended arching effects often prove disastrous in industrial applications, as we will see later in the case of grain silos.

Aleatory Architectures

Aleatory geometrical structures like those in figure 7.6 have provided a source of inspiration for contemporary architects. The side walls of the recently constructed Beijing National Stadium comprise an open structure made from steel bars attached to contacts in apparent disorder. The stadium's nickname, "Bird's Nest," refers to the random tangle of twigs and sticks in a well-built structure made by feathered creatures. Networks of force lines in granular models inspired the rise of such designs. Arrangements like this offer advantages that go beyond aesthetic considerations inasmuch as they prove to be lightweight and open, distributing stress and strain in a way that corresponds to the packing of well-fitted particles on a smaller scale.

PRESSURE UNDER A PILE OF SAND

The micro-arches described previously are small structures within a granular medium. A micro-arch can involve a certain number of grains; yet its size is small with respect to that of a pile or a silo containing millions and billions of grains. How, then, are micro-arches able to influence the behavior of a granular material on a large scale? As we have noted, reduced packing fraction is a consequence of arching. The volume of salt we pour into a shaker can diminish by 10%, while shaking will destroy micro-arches contained in it; this represents a macroscopic arching effect. By the same token, the redirection of forces within a granular medium can also occur through arching on a microlevel. Thus, the vertical stress due to the weight of grains is transmitted toward the bottom of the shaker and its sides: granular material in a vertical container exerts a horizontal pressure on the side walls. In contrast to liquids at rest, where pressure is the same in all directions, the pressure that a granular medium exerts on the container's vertical surfaces can be greater or smaller than the pressure exerted on a horizontal plane. A simple way to explain this difference is to say that it depends on the orientations of micro-arches or force chains. A classic problem in soil mechanics is calculating the pressure that the soil in question exerts on supporting walls. At a given depth, the pressure on the wall can vary between one-third and two-thirds times the weight of grains at the same depth, depending on the deformations experienced by the soil and, therefore, its texture.

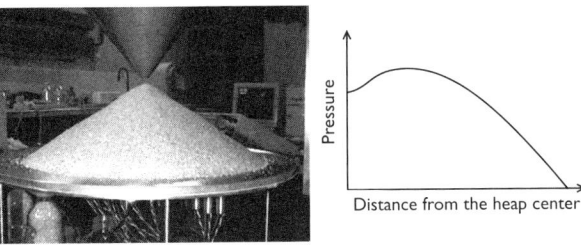

Figure 7.7
A pile of sand obtained by pouring grains from a funnel. The pressure measured at the base of the pile, as a function of distance to the center, reaches a local minimum value at the vertical of the top of the pile—which is paradoxical.

An example illustrating this redirection of forces is the distribution of pressure under a conical pile of sand. It is well known that hydrostatic pressure is proportional to depth below the free surface in a liquid. Consequently, one might think that the pressure a pile of sand exerts on its base is maximal at a vertical direction from its apex. This is not the case, however. If the pressure is measured from the edges of the pile toward the center, we can observe that pressure does indeed increase, but in a nonlinear fashion. Even more surprisingly, we see that pressure reaches a local minimum at the vertical of the center of the pile (see figure 7.7). The presence of this minimum value surprised researchers when they performed highly precise pressure measurements. In this way, they showed that the pressure value at the center depends, in part, on the method employed to construct the pile. Such variability calls to mind that of the stress on a supporting wall, which depends on deformations the soil has undergone.

Why isn't pressure proportional to the height of sand columns right at the top? Simply put, it's because the forces within the column, which result from the weight of grains, are not oriented purely downward. Each column is supported by the column "downstream" and supports the column "upstream." Without getting into all the details, we can say that the column right at the middle of the pile is supported by other columns, and therefore that part of its weight is supported by all the grains found in its vicinity. The situation resembles a row of books slightly leaning against each other; if you lift a book stuck between two others and then let it go, it will remain suspended because of the forces of friction exerted by its neighbors. In other words, its weight is balanced by forces of friction; the book exerts no force of its own on its base (i.e., the shelf).

As such, a pile of sand may be described as a cathedral made up of countless micro-arches. The overall result is a macroscopic or bulk arching effect,. These arches reflect friction between grains and lead to the existence of a slope angle.

PRESSURE IN A SILO

A funnel filled with rice or, on a larger scale, a silo filled with grains can become blocked; a knock on the side, and the flow starts up again. Let's take a look at a cylindrical silo (figure 7.8a). In the absence of friction between the grains and side walls, the pressure at the level of its lower opening would vary in proportion to the height, h, of the contents (represented by the straight line, 2). In fact, however, friction between grains and the side walls acts on the particles and works against their downward movement (figure 7.8b). Because of this friction, part of the grains' weight is supported by the walls of the silo; the force F exerted at its bottom (figure 7.8; curve 1) increases more slowly with height than if there were no

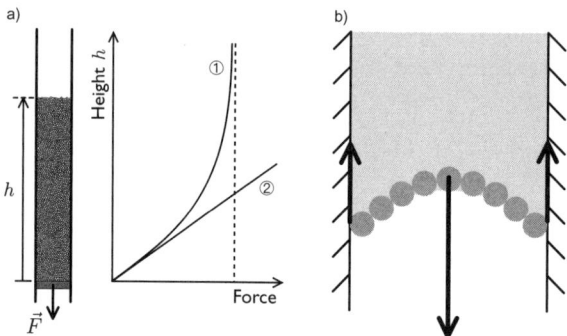

Figure 7.8
a) Illustration of the Janssen effect in a silo. The force, F, applied to the base of the column is in large part supported by forces of friction exerted on the side walls. Increase of F with the filling height h is slower without wall friction (curve 1) than with wall friction (curve 2). b) Schematic representation of the blockage of the silo flow. The vertical force due to the weight above is, to a large extent, supported by forces of friction mobilized along the walls.

friction (curve 2). In particular, beyond a certain level, F practically no longer increases; all the extra weight, if one keeps on filling, is supported by the walls. This is known as the Janssen effect, named after the German physicist who first described it at the end of the nineteenth century. It is as if the the granular column were hanging from the silo's walls for the most part. The situation is quite similar to the position assumed by a climber between two opposing walls stretching out his arms and legs and leaning against both of them to support his weight.

The Janssen effect is present in grain flows in a tube or between two parallel planes: for example, ballast particles (presented in chapter 2) rest on a previously compacted soil and are partially confined by railway ties. Under the weight of passing trains, the particles are compressed between the tracks and the ground, making them stretch horizontally. Since the lateral reaction pressure exerted by the soil is weak, the spreading of the ballasted layer is prevented by the horizontal friction forces between the ballast particles and the ties.

Another example of this effect is the friction force that granular material exerts on an object that has been stuck into it. This is easily observed by means of a simple experiment. Place a stick vertically in the middle of a cylindrical bucket and pour in grains, up to the top. Now, the piece of wood is stuck in granular matter. Pull straight up, and the stick will come out, in spite of the friction the grains exert on it. But if one densifies the material by tapping on the sides of the bucket after it is filled, the stick won't come out. In fact, it will even lift the container up! This dramatic demonstration shows that increased packing fraction reinforces the Janssen effect.

These two last chapters have shown us the heterogeneity of granular media. Whereas the organization of fitted stones ensures stability in the vaults of cathedrals, granular media consist of grains in disorder. A pile of sand—a messy heap—is proverbial for being impossible to describe in detail. That said, and counter to all expectation, the average effect of such disorder can be reproduced and even controlled by applying stress to the grains.

The percolation model offered us a rigorous scheme that applies to electrical properties, which are directly tied to the organization of the most stressed contacts and the force chains between poorly conducting grains they produce. In contrast, in order to describe a medium's mechanical behavior, one must also consider the weakest contacts, which prop up the strong lines of force, like a flying buttress, and prevent them from giving way. Resistance to deformation is a consequence of the very large number of small forces and the much smaller number of large ones. The vectorial character of the latter and of the contact network means that an anisotropic organization ensures overall balance. The weak link is resistance to friction between grains. If friction is reduced, the probability that arches will form diminishes, and therefore, mechanical strength decreases. All the same, strength cannot be increased indefinitely in this manner, because large-scale disorder doesn't permit the formation of macroscopic arches (or arranging well-fitted stones as in an arch). Each grain has to find its place. As interactions proceed at random, the granular medium assumes order by forming, on the various scales, load-bearing structures of its own made up of linked grains. In this context, then, Victor Hugo's prophetic words in *Les Misérables* provide an apt conclusion: "The small is great, and the great is small; all is balanced in necessity."

8

GRANULAR FLOWS

It is smooth and round atoms that form bodies whose substance is liquid and fluid. For atoms of spherical form cannot remain unified; when spilled, they roll easily, as if downhill.

Lucretius, *De Rerum Natura*

Screes, landslides, and avalanches are examples in nature—often spectacular and disastrous—of grains of matter in collective motion. Dunes that move over vast distances shape the desert landscape. Coasts take form through sand being conveyed by marine currents. Crushed ore is convected by air flux in tubes over long distances. The processing of grains in agriculture, from harvest time to when they are distributed from storage silos, calls liquid flows to mind.

All of these phenomena involve subjecting grains to powerful agitation; they contrast with situations where deformation occurs slowly (which we have already discussed). Whereas Coulomb's law of friction accounts for slow flows, in order to explain the irreversible process of plasticity, when rapid flows occur, we must also consider *inelastic collision** between grains.

In this dynamic regime, grains undergo acceleration: their kinetic energy and inertia (i.e., resistance to changes of speed) modify the properties of the flow, that is, of its *rheology*.* In such a case, one speaks of a regime of inertial flow. The distinction is the same as the one we make, apropos of a liquid or gas, between a viscous regime (when the flow is sluggish) and turbulent and inertial flows that are rapid. The *Reynolds number*, which will be presented in chapter 11, characterizes the domains of existence of these two regimes. By the same token, for understanding the case of slow and inertial grain flows, a dimensionless number specific to granular flows proves useful. This *inertial number** plays a role analogous to the role of the Reynolds number for fluids.

This chapter examines both regimes. We will first consider steady flows, then transient ones, such as an avalanche of grains down a slope. We will also consider the case when applied vibrations underlie the dynamics of grains.

GRAIN INERTIA

In a rugby match, a team's advance toward the goal line can be slowed down by interference or a melee with the player carrying the ball, to say nothing of a tackle. Likewise, in a granular medium, flow can be braked and deviated by friction or shocks between grains. When the shear rate increases, the shocks increase in number and amplitude, leading to enhanced resistance. It's a bit like what happens when you try to move in a dense crowd: in contrast to an organized parade, where everyone marches at the same rhythm and in the same direction, it's difficult to move in an open market while avoiding other people; collisions prove inevitable. Experiments have shown that deceleration is proportional to the number of shocks that occur. In this light, it can be demonstrated that everything happens as if each particle were subjected to a force proportional to the square of shear rate.

Consider a granular material confined between two parallel plates, one fixed and one moving at a velocity V, while their distance of separation h remains constant (figure 8.1). This configuration is called *simple shear,** and it applies to liquids as well as to diluted suspensions of grains. The quantity that measures the intensity of shearing is not the relative velocity V of the two plates, but this velocity normalized by the thickness h of the granular layer. This ratio $\gamma = V / h$ is called *shear rate* and has the unit of the inverse of a time. If flow is homogeneous, the relative speed,

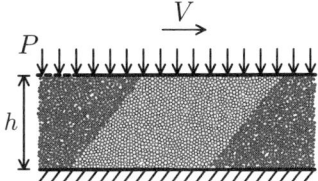

Figure 8.1
Homogeneous grain flow between two planes. The upper plane moves at velocity V with respect to the bottom plane. A pressure P is applied on the upper plane. The volume can change when the granular matter dilates under the effect of shearing. This configuration is often used in soil mechanics.

v, between two grains with centers separated by height d (corresponding to an average diameter) is $v = \gamma d$; it is the order of magnitude of the relative velocity between two grains in frictional contact in two neighboring layers. A basic law of dynamics, Newton's second law, states that the force between grains should be balanced by the product of a mass multiplied by acceleration, which corresponds to the impulse mv exchanged per unit of time between grains of mass m. During shearing, this quantity is $mv = m\gamma d$, the characteristic time of the exchange corresponding to the inverse of the shear rate. The ratio of $m\gamma d$ and $1/\gamma$ yields the inertial force $F_i = md\gamma^2$ acting between grains. This force, the intensity of which is proportional to the average mass m and the square of the shear rate, represents a dynamic effect.

Before proceeding, it is important to distinguish shearing under constant pressure (as presented in figure 8.1) from shearing at constant volume, or simple shear, for a liquid or a suspension. In nature and many industrial procedures employing grains, volume is not fixed, and the granular material is free to expand or contract. We saw in chapter 4 that this effect of dilatancy is a general property of granular materials. When grains' packing fraction is larger than the critical packing fraction, a preliminary dilation is necessary for shear deformation to occur. Under such conditions, if the distance h between plates undergoing shear remained fixed, the granular material would be blocked. In order to avoid blockage, a pressure P is applied, which confines the material while allowing h to vary during shear flow. This situation isn't very different from the case presented by flow down an inclined plane, in which pressure comes

Inertial Number

Consider the configuration of a simple shear with imposed pressure, as in figure 8.1. The average static force for each grain, $F_s = Pd^2$, is due to the confining pressure P. This force may be compared to the inertial force $F_i = md\dot{\gamma}^2$ that comes from collisions, as seen earlier. The square root of the ratio between these two forces is a dimensionless number called the inertial number:

$$I = \sqrt{F_i / F_s} = \dot{\gamma}\sqrt{m / Pd}.$$

Hereby, using the square root enables one to obtain a number proportional to the shear rate $\dot{\gamma}$, so that the inertial number may also be viewed as a "reduced" shear rate.

from the mass m of the layers of grains: its average value is the weight mg divided by the average grain section d^2.

The inertial number I is small for a low shear rate, if confining pressure P is high, or if the mass of a grain is small. When I has a low value—typically, below 10^{-3}—the inertial effects of grains are negligible. In such cases, one speaks of quasi-static or slow flows. In the opposite case, flow is described as inertial.

FLOW REGIMES

The inertial number I allows one to unify the rheology of different grain flows by means of a law describing the full range of their possible behaviors. This is what a research group called MiDi (*Milieux Divisés*, "Divided Media") did in 2004 at the National Center for Scientific Research (CNRS) in France. Researchers compared grain flows in several experimental arrangements: cases of plane shear between two plates; in silos; on inclined surfaces; in a rotating drum. They measured the speeds of grains by using photographic images in order to examine collections of grains as well as individual marked grains. By this means, variations of packing were determined, as well. The behavior for various geometrical configurations boils down to two laws involving the inertial number I, as represented in figure 8.2. Measurements show (figure 8.2a) that the larger I is, the more the packing fraction is reduced; in other words, the sample dilates and takes up more space. Figure 8.2b shows the evolution

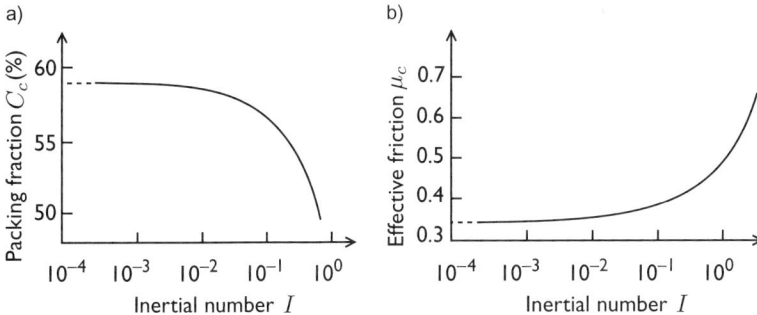

Figure 8.2
a) The increase of the inertial number I is accompanied by increased collision effects, resulting in a decrease of packing fraction C_c (critical packing fraction during flow). b) The dissipation of energy, which is no longer due simply to friction but also to the inertial effects of collisions, increases, leading to the increase of the effective coefficient of friction μ_c.

of the effective coefficient of friction μ_c as a function of I (figure 8.2b). Hereby, μ_c (the ratio τ/P) increases with I, which indicates that as speed increases, it is more and more difficult to shear a granular material as a result of the increasing number of collisions between grains.

When the inertial number is close to zero, static forces F_s between grains prevail. At this quasi-static limit, the coefficient of friction $\mu_c = \mu_0$ corresponds to the effective coefficient of friction (see figure 5.9). The minimum shear stress τ_0 required to put the granular material in motion is $\tau_0 = \mu_0 P$, and the packing fraction reaches its critical value C_0, as shown in figure 4.7.

When the inertial number increases, the dissipation of energy due to inertial forces in collisions adds up to the energy loss due to friction. The overall friction μ_c increases from μ_0 and the difference $\tau_c - \tau_0 = P(\mu_c - \mu_0)$ reflects the dissipation of excess energy due to inelastic collisions between grains. This range of intermediate values of I is also called a *liquid-like regime*, analogous to viscous molecular liquids in which shear stress increases with shear rate. If the distance between plates remained constant, the granular matter would not change volume, and its packing fraction C would remain fixed, whereas pressure P would increase with the shear rate.

Experiments and simulations show that τ and P are proportional to γ^2, with proportionality constants dependent only on the packing fraction. This law is known as the Bagnold formula. Ralph Bagnold was a superior officer in the British army serving in the Libyan desert in 1935. He described the effects produced by wind in the formation of dunes. His knowledge of the terrain and study of possibilities for military vehicles to move on sand played a vital role until combat against the Germans and Italians ceased in 1943. Science is indebted to Bagnold for being the first to appreciate and describe the importance of collisions between grains for overall displacement. His point of departure was the hypothesis that grains are in a *gas-like* state and interact only when they collide. This work laid the foundation for our modern understanding of grain flow. On this basis, scientists have elaborated a kinetic theory of granular motion, employing analogies between the ways that grains and molecules move. The gas regime corresponds to high values of the inertial number at constant pressure, or to small values of the packing fraction for volume-controlled shear. Thus, Bagnold's formula also applies to a *liquid-like* regime, even if interactions here arise both from lasting contacts and from collisions between grains.

The analogy between a rapid granular flow and a molecular gas for large values of the inertial number is inspired by the thermal agitation of small grains or molecules. All the same, two major differences exist.

The first difference concerns *thermal agitation,* a manifestation of the absolute temperature of a gas. Each particle draws energy from thermal agitation (see the inset in chapter 1) measured as the product of the *Boltzmann constant* k_B and the absolute temperature T (300 K at room temperature). This energy $k_B T$ takes the form of kinetic energy $\frac{1}{2}mv^2$, where m is the particle's mass and v^2 the mean square of agitation speed in all directions. At room temperature, the speed of gas molecules amounts to a few hundred meters per second. However, for a grain 10 μm in diameter, the volume of which is 10^{14} larger than an atom of oxygen, agitation speed would be just one-millionth of the speed of a gas molecule. Accordingly, the agitation of particles of snow on the surface of an avalanche does not come from Brownian motion, but from nudges and bumps that occur as the top layer tumbles down the slope, which maintains the particles' relative motion as the ensemble flows down. Scientists still do not know exactly how to establish a relationship between *granular temperature* (which may be understood in reference to thermal agitation) and the mechanisms that give rise to it.

The second major difference concerns the nature of collisions. For a gas, these shocks are *elastic*: the total kinetic energy of two molecules striking each other is conserved during the shock. No energy is lost, and molecules' motions amount to perpetual motion! But for grains of matter, part of the kinetic energy vanishes in the form of heat in each collision; this heat may be high enough to bring particles of dust to incandescence. The coefficient of restitution e which has a value between 0 and 1, measures the rebound efficiency of particles and reflects their degree of inelasticity (chapter 5). In practice, the model of kinetic theory can be applied accurately only if this coefficient is sufficiently close to 1.

The conditions of thermal agitation of gas molecules may be approached by making particles vibrate while reducing friction. In teaching laboratories, an air table is used: a horizontal surface dotted with small holes through which puffs of air keep pucks afloat, with minimal friction. The same principle is at work in Spacetrains (air-cushioned vehicles)—except that in our experimental work we expect pucks to collide! A good model for thermal agitation of a gas can be obtained by making the sides of the table vibrate slightly to encourage pucks that have reached the edges to keep moving and retain their agitated state.

This chapter started with a look at the shear strength of an assembly of grains placed between two parallel plates subjected to relative motion. This provides a remarkable example of homogeneous flows. In practice, however, the flows encountered in nature and common procedures are rarely homogeneous or in a steady state. The rest of the chapter will examine the following cases: hourglasses, avalanches on a granular slope, instabilities of vibrated piles, and an array of effects produced by mixtures of grains (intriguingly known as the "Brazil nut effect").

HOURGLASS AND SILO

The classic timer for making the perfect softboiled egg—a mini hourglass holding just enough sand for a three-minute countdown—has become scarce in the last few years. And yet, what a fine object for study! For some seven centuries, the traditional hourglass was used for measuring the course of time, especially on voyages across oceans; in antiquity, only the clepsydra—or water-clock, by which time was measured by a continuous flow of liquid from an opening—was known.

When a trickle of water passes through a small opening in a water clock, the flow rate diminishes as the water level in the upper chamber decreases. But in an hourglass, the flow rate remains constant until the top half is almost empty. That's the first paradox. A key factor is the friction of grains on the sides of the vessel, which works against the transmission of forces toward the bottom of the pile. Let's revisit the experiment for visualizing stresses by means of photoelasticity (introduced in chapter 7): the transmission of stress occurs via a fairly complex network of forces. The grains, under the weight of higher layers of grains, form continuous arches, below which many grains are practically free (figure 8.3). This is the reason why the grains at the level of the opening don't need to support the weight of all the grains on top of them; the flow is independent of the amount of sand above the opening (figure 10.3).

We can understand the formation of arches at the opening between the chambers of the hourglass with the analogy of a fire in a theater: If the audience panics, everyone will rush to the emergency exit. The flow of people who manage to get out will not be a function of those still in

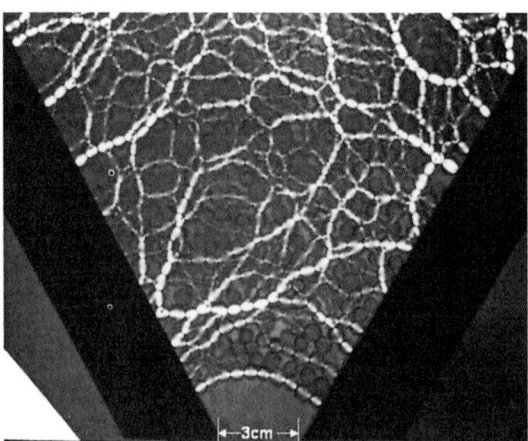

Figure 8.3
Snapshot of a photoelastic grain flow in two dimensions. The vaults that form redistribute the weight of upper layers to the side walls.

the room, but of those who try to overcome the bottleneck at the door. Indeed, models of granular material are now being used to regulate traffic, especially jamming, and situations of collective panic. Paradoxically, it's sometimes helpful to partially block or limit circulation temporarily in order to increase the evacuation rate. Models employing grains or pucks on a conveyor belt allow such conditions on a human scale to be simulated.

So how does flow depend on the radius of a circular opening with diameter D? In the case of a liquid, the flow rate is proportional to the area of the opening $\pi D^2 / 4$ and to the overpressure resulting from the water level in the vessel. For grains, however, many experiments have shown that the flow rate varies as $(D-d)^{5/2}$, where d is the diameter of grains. This law, which was discovered by Gotthilf Hagen in 1852 and experimentally proven by W. A. Beverloo in 1961, follows from the presence of arches above the opening. When exiting, the grains fall at a speed greater than zero—indicating that they have undergone free fall inside the silo just above the opening. In keeping with the principles outlined here, the height from which the grains have accelerated is of the same order as the diameter of the opening. For uniformly accelerated motion under gravity, the velocity attained from a height equal to D varies as $\sqrt{D}$.

To obtain the flow rate—in other words, the number of grains leaving the silo per unit time—the number of grains per volume unit in the cross section of the opening (which is proportional to D^2) is multiplied by the velocity, yielding a dependence of $D^{5/2}$. That said, note that the part of the opening that proves "effective" for the passage of grains is not D, but $D-d$; a hole diameter of the order of d wouldn't allow any grain to pass.

The flow in an hourglass is relatively stable only if the diameter of the opening between chambers is much larger than the diameters of grains. For spherical grains of the same size and without cohesion, the diameter of the opening should be at least six times that of grains. This limit can increase considerably in the event of irregular grain forms, significant friction, forces of adhesion when grains are sufficiently small, or capillary forces for a humid medium. Recent studies indicate that when the diameter of the opening is below this limit, the flow stops after a finite time, and the number of grains that pass before the flow stops follows a *Poisson process*, named after the French mathematician Denis Poisson. "Process" refers to the occurrence of a series of infrequent events over time. For instance, births may be studied as a function of time; a Poisson process is at work inasmuch as the birth of one child occurs independently of other preceding births. Another term is "memoryless process."

Now consider flows in a silo, which depend on factors related to the material itself (grain geometry and the nature of contacts between particles, which vary with humidity), the geometry of the reservoir (aperture angle and diameter), and interactions between the grains and the reservoir (friction at the walls). As in the textbook example of the hourglass, flow can be interrupted when a stable arch forms over the opening. The situations encountered when a silo is being emptied are, in fact, highly diverse and still not well understood. The poor control we have over grain flow is evident in the number of bumps on the sides of metal silos: one good way to get things moving again is to give the outside a hard knock!

In general, measures are taken so that, on average, all the grains will move toward the opening in a uniform fashion, whatever their distance from the axis of the silo may be—which is called "mass flow" (figure 8.4a). This occurs if the silo is relatively narrow and has a low aperture angle. In the opposite case, a different state is obtained where only grains along the axial portion flow, and they slowly draw along particles on the upper surface of the material (figure 8.4b). This is known as "funnel flow." If the slope of the convergent part of the silo isn't big enough, or if major friction

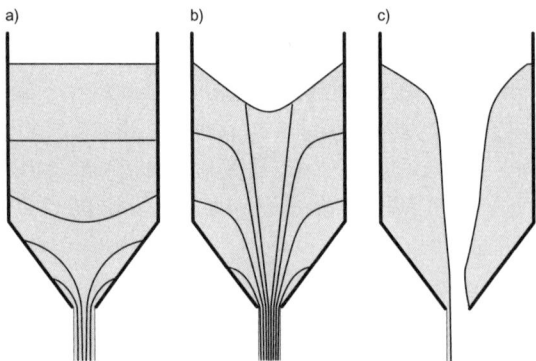

Figure 8.4
Different flow modes in a silo: a) mass flow; b) funnel flow; c) accidental formation of a "rat hole."

exists, the flow may stop before draining is complete; in extreme cases, a pathway known as a "rat hole," extending to the upper level of the grain mass, will form (figure 8.4c).

When a chimney or an arch ruptures, whether spontaneously or because of action from outside, one possible result is a "run," or, in other words, an abrupt torrent of the material. In falling, the material captures a significant amount of air as it comes into contact with machinery (for instance, a conveyor belt). In turn, it will set this air free at a speed that depends on its permeability: for large grains, deaeration occurs quite rapidly; if grains are small, the process is much slower. In the second case, the captured air is under high pressure. As it escapes, it fluidizes the material, which now flows as if it were a liquid—potentially overflowing and spreading over a sizeable expanse. In a factory, significant damage may result.

SLOPE ANGLE

When a pile is made by adding sand to a substrate at a steady rate (as occurs in the lower part of an hourglass), the result is a cone with an angle to the summit that remains constant over time; this feature distinguishes it from a liquid. As the sandpile grows, it preserves the same angle by allowing extra grains to roll down the sides. The angle the cone forms on the horizontal plane, when the addition of grains stops, is called the "angle of repose," or "slope angle" (cf. chapter 5); its notation is θ_r. This value depends on the surface state and shape of grains, boundary conditions, and so on.

The equilibrium conditions for sand piles have provided the focus of many recent studies bearing on problems of critical state (see figure 4.7). One simple experiment uses a transparent cylinder partially filled with spheres, which rotates around a horizontal axis (figure 8.5). This arrangement allows the slope angle to be varied progressively, and it also permits grains to be added continuously at the top of the pile. Initially and at a low speed of rotation, the granular medium follows the cylinder's rotating motion as a whole, like a solid, provided that the angle of its free surface on the horizontal plane remains below θ_m; this is the same as the maximum angle of stability θ_p (see chapter 5)—and lies a few degrees higher

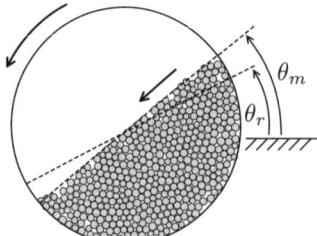

Figure 8.5
Maximum angle, θ_m, and angle of repose, θ_r, of a pile being turned in a cylindrical container.

than the angle of repose θ_r. When this value θ_m is reached, avalanches begin at the pile's surface and quickly bring the angle back to its repose value. Then, the avalanches stop, the angle rises again, and so on. This experiment shows that two angles are necessary for defining the stability of a granular medium. Between the values θ_r and θ_m, the situation is complex; the pile can be stable or evolve, depending on the way it is made and has been subjected to perturbation (as we have seen with regard to ski slopes). This sensitivity to avalanche effects can be made to vanish by using nonconvex particles (e.g., cross shapes), as discussed in chapter 4.

The Coulomb law of friction, which relies on a single critical angle, is not enough to understand such behavior; the Reynolds model, which we introduced to account for the phenomenon of dilatancy (figure 4.8), must be employed. Osborne Reynolds studied how the physical properties of grains influence the coefficient of solid friction when a container holding sand is tilted progressively. The difference between the maximum angle of stability θ_m (for setting the sand into motion), and the angle of repose θ_r derives from the fact that grains of sand subjected to shear along a given direction must be set free from their "cage" of other close grains hindering their motion. This implies that they should move transversally with respect to the flow direction over a finite distance δ. This transverse motion reflects the dilatancy, which can be expressed as a transverse deformation δ / d, where d is mean grain diameter. The transverse deformation is generally small and can be approximated by an angle $\alpha = \delta / d$, corresponding to the approximate value of the difference $\theta_m - \theta_r$ between the maximum angle of stability and angle of repose.

VIBRATED PILES

Research on avalanches has enabled scientists to understand a phenomenon Michael Faraday described more than a century ago. By vertically vibrating a plate holding a horizontal layer of powder of sufficient amplitude, one observes that the surface of this layer becomes unstable and forms regular peaked geometrical structures such as squares and triangles. The phenomenon also occurs when vertical vibration is applied to a liquid layer; it might seem tempting to describe it along the same lines.

If a single grain is placed on a horizontal membrane vibrating vertically, at amplitude a and frequency ω, it will remain fixed as long as the vertical acceleration—proportional to a and to the square of the frequency—that the membrane transmits to the grain is below the acceleration of gravity, g, or, in other words, as long as the gravitational force (the weight) is greater than the acceleration force coming from the membrane's vertical movement. When $a\omega^2$ exceeds g in the course of upward movement, the particle will detach from the membrane and begin to move freely under the effect of gravity and its own speed, as if on a trampoline. It proves especially interesting when the particle strikes the membrane again as the latter is in the same phase of movement, after completing a full period. Now, the particle benefits in optimal fashion from the impulsion provided by the membrane. Under these conditions, resonance occurs. In general, however, a particle will fall at a different phase; the motion that results from this "trampolining" exemplifies a chaotic phenomenon.

The same phenomena are observed in a packing of spheres. If vertical acceleration exceeds a threshold value, the particles start to move, and the surface becomes unstable. A pile forms, fed at the top by the spheres' convective motion under the influence of the container's sides. The avalanches that occur on the pile's surface counterbalance this internal motion. The effect proves quite significant for fine powders (grains of diameter below fifty microns), which rapidly fluidize. The phenomenon may be compared to the formation of cellular structures in a layer of liquid heated from below, arising from the coexistence of the upward flow of hot liquid and the downward flow of cold liquid.

SEGREGATION

Grains do things their own way … yet again! If an ordinary fluid comprising different phases is shaken or vibrated, the operation will help to mix the whole. For a mixture of dry grains, on the other hand, the same operation can lead to unmixing, depending on the elements' geometry and other physical characteristics. This process often proves desirable for sorting grains according to size. Most commonly, however, a uniform composition of grains is sought.

BRAZIL NUTS

This section heading refers to an observation made in a letter to *Physical Review* published in 1987—"Why the Brazil nuts are on top"—that provoked immediate interest among physicists for a matter already known to researchers of soil mechanics and powder technology! Brazil nuts are large grains, a few centimeters in diameter. When piled in trucks mixed along with other smaller nuts, they are found on top after a trip down long, bumpy roads. This phenomenon represents a particular case of the segregation that occurs in mixtures of grains.

Several mechanisms may be invoked to explain the effect. The first involves the fact that, during the upward phase of motion of vibrating spheres, the spheres detach from each other—as just described in the context of a vibrating bed, when vertical acceleration is stronger than that of gravity. Thus, the small particles have an easier time sneaking into the spaces freed up below larger ones, and they fill the lower part of the vessel bit by bit. A second mechanism involves the overall convective motion of vibrated spheres, which carries big and small spheres alike up to the surface; the big grains remain lodged here, whereas small spheres move down the vertical sides of the vessel due to their size. Disagreement among researchers concerning the relative importance of these two mechanisms proves that, in contrast to fluid mechanics, no unified science of grain flows exists to date. For the case at hand, however, experiments in two dimensions have allowed the situation to be clarified: geometric effects are most active at a low level of vibration; when vibration is strong, convection along the sides prevails.

To understand segregation, it is necessary to understand how a mixture of grains of unequal sizes behaves. When such a mixture sediments in a liquid, the most natural distribution for large grains is to fall to the bottom at a higher rate than small ones. The result is a normal gradation such as one finds in sedimentary rocks and deposits at the bottom of oceans, where the size of particles decreases from the bottom to the top of the sediment. On the other hand, dry granular media present an opposite gradation; here, the large grains are on top. Look at a package of cornflakes, and you'll see the small broken bits at the bottom. Likewise, plowing a field will bring new stones to the surface year after year, as smaller grains are buried beneath large ones little by little.

Segregation may serve to separate materials according to size. That said, most applications of granular materials aim for mixtures that will be as homogeneous as possible. Just imagine the consequences that a poor mixture of components of concrete (cement, aggregate, additives) can have on the mechanical strength of a structure. By the same token, try to picture pharmaceutical pills made simply by compounding powder in a container with elements that separate at various filling levels. In this light, it's easy to appreciate the interest the subject raises in all areas where grains represent the base material. To avoid difficulties, such mixtures are often prepared by means of wet milling, which limits segregation.

Segregation can occur with differences of density, shape, surface state, or size between grains; size difference is by far the most important. In fact, segregation is present each time that the "history" of a system involves relative motion of different grain types. For the sake of simplification, suppose we have just two kinds of spheres, with different diameters. In a pile of spheres with the same diameter d, the smallest gaps will be curvilinear triangles formed by three large spheres in mutual contact. A small sphere will be able to pass through this space, if its diameter is below 0.15 d. Small spheres will manage to circulate freely through pores in the pile of large spheres, even if the latter are at rest, because of the action of forces of gravity. If the difference of grain sizes is lower, the infiltration of smaller grains into the pores between larger ones will not be spontaneous; it will occur, however, if the material is set in motion.

SEGREGATION IN AVALANCHES

In a mountain corridor formed by a rockslide, one can observe that, on average, rocks get bigger and bigger the farther down they are. On account of their weight, the smaller grains on the surface roll along with big ones in a scree, but they readily fall to a lower layer of grains if, in the course of moving, they happen to encounter a space of suitable size to accommodate them: the top layer of the avalanche acts like a sieve; only the smallest grains pass through. Large grains remain on the surface, and they are also less sensitive to its roughness; because of their higher inertia, they travel a longer distance by rolling and bouncing. Their agitation along the slope has forced smaller grains to become buried, as if the dynamic sieve were being shaken regularly.

Our world offers numerous examples of grain transports of this kind. Long ago, in prehistoric times, it seems that a landslide in Flims (Switzerland) set a volume of several cubic kilometers into motion for a few minutes. The energy expended would have matched worldwide energy consumption for half a day! Unfortunately, such cases are not unique to prehistoric times. Avalanches and screes inspire fascination or terror, in keeping with the distance at which they are observed. Winter sports technology along with the disappearance of traditional activities (agriculture and pasturing) favor the increase of such geological catastrophes and make it more necessary than ever to understand how they occur.

For most cases in the natural world on a grand scale, it is necessary to consider the presence of a fluid, in addition to the mechanisms of friction and collision we have been discussing for dry grains. The following chapters will explore the presence of water between grains. In chapter 9, we will see how the presence of water creates adhesion effects—as in a sandcastle. The case of permanent adhesion will be discussed in chapter 10. Chapter 11 examines the relative flow of a fluid inside a dense granular medium. The last chapter, chapter 12, will explore joint grain-liquid flow, as in the cases of avalanches or screes of torrential muds.

9

FROM SANDCASTLES TO CLAY TOWERS

History is a child building a sandcastle by the sea, and the child is the whole majesty of man's power in the world.

Heraclitus

Building sandcastles is a popular activity, sure to garner attention. Recently, an exhibition at the Pompidou Art Center in Paris for young visitors was devoted to this form of architecture. A television show in France also organized a contest to build a sandcastle big enough for a person to step inside (an author of this book monitored the challenge!). A vast scientific community has taken interest in the strength of structures of this kind: it concerns effects of cohesion and packing fraction, which depend on arches stabilized by capillary action and the presence of clayey soil. The matter is of even greater import inasmuch as a sizeable portion of humanity still uses and dwells in earthen structures.

FLUID BETWEEN GRAINS

Until now, we haven't considered the elements between grains in a packing: air in the case of coarse grains in a pile, water in a saturated granular soil, and both air and water in unsaturated porous media (figure 9.1). Depending on the level of humidity or dryness, a granular material can

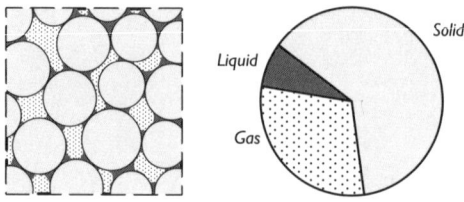

Figure 9.1
Wetted sand is a triphasic material composed of grains, water, and air.

be in any one of these states. That said, water doesn't just make its way between grains: through interaction with solid and gaseous phases, it creates so-called *capillary* bonds, which make particles stick to each other, lending mechanical strength, or cohesion, to the material. Chapter 10 will examine other mechanisms of cohesion. These capillary forces play a decisive role when building a sandcastle, for instance. Moreover, the presence of fluid between grains may favor small displacements and entail changes of packing fraction.

In a porous medium, other fluids can occupy the spaces left free. This is the unfortunate case of soils polluted by oil deposits—or, under more favorable conditions, when hydrocarbons are extracted from source rocks; here, oils, tars, and any number of admixtures share the space left free with air. Chapter 11 will turn to flows in which several fluid phases are present in a porous medium. For now, we will examine the case of air and water in contact with grains simultaneously.

SURFACE TENSION

A molecule within a liquid at rest is subject to forces of attraction exerted by all the molecules surrounding it; the resulting force is zero on average. However, a molecule on the surface is drawn toward the interior of the liquid (figure 9.2). Consequently, the molecules at the interface between a gas, a solid, or another immiscible liquid have a higher energy level than the molecules within the liquid. Such excess energy per surface unit is called *surface energy*. The free surface tends to minimize this energy and therefore to take up as little area as possible—provided that one takes into account external forces to which the liquid is subject, such as its weight.

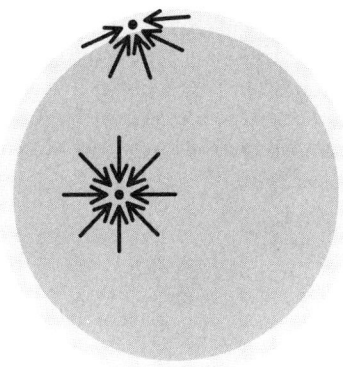

Figure 9.2

Attraction forces between molecules at the interface between liquid and gas exert a tensile force that bends the drop's surface.

In turn, this minimization translates into a force per unit length of the interface (or an energy per unit area): its *surface tension*.* The *Laplace-Young law** connects the curvature of the interface to surface tension and to the pressure difference between the two sides of the surface (see inset following).

In the absence of an external force, a liquid will assume a spherical form: a given volume of liquid displays as small an outer surface as possible. A tiny drop of rain or fog remains spherical: if drops of water in the air deviated from spherical form just a little bit, there would be no rainbows—which are due to multiple reflections of sunlight in spherical drops! And if a scientist like Captain Haddock in the *Tintin* series tried to pour a glass of whiskey in a zero gravity flight, the liquid would assume the shape of a spherical glob floating in space.

Laplace-Young Equation

In order to minimize surface area, a small drop of liquid assumes a spherical shape. The energy per surface unit area, which corresponds to a force per unit of length, is its surface tension γ_s. As with a rubber balloon, the pressure inside the bubble is higher than that of the atmosphere. Just as a spherical balloon resists being inflated—especially when it's small—the overpressure varies as the inverse of the sphere's radius r. In brief, the *Laplace pressure* is expressed as $\Delta p = 2\gamma_s / r$. In more general terms, which apply to the surface of nonspherical drops, the Laplace-Young equation relates the overpressure to the curvature C as $\Delta p = \gamma_s C$. The *mean curvature** C is determined on the basis of two principal radii of curvature, $C = 1/r_1 + 1/r_2$. For spherical gains where the two radii are equal to r, one recovers the previous expression. For a meniscus between two grains, the two radii of curvature lie in opposite directions (figure 9.3).

THE PHYSICS OF SANDCASTLES

Surface tension allows us to account for the way a liquid "gets along," for better or worse, with a solid with which it is in contact. In the first instance—and this is usually the case for sand (silica) and water—the liquid wets the grain surface, as it would also do on a clean plate of glass; the solid is *hydrophilic* (water wets glass easily). However, it is also possible by chemical means to attach to the surface an organic compound that doesn't like water, so that water cannot wet the surface of the grains. Doing so yields a *hydrophobic* sand, with which it's even possible to make a castle underwater: the air that surrounds each of these grains chases the water away, and the air keeps the grains under pressure. "Magic sand" of this kind can be bought in toy stores. If it weren't for the cost, such sand could be pulverized and scattered on beaches polluted by tar; its oils would seep in between the grains, as water does in ordinary sand.

WET GRAINS

The grains of a sandcastle on the beach are wetted; this is what makes it possible to construct the castle in the first place. A given drop of liquid no longer has a spherical form when it connects two grains. Consider the drop represented in figure 9.3, which is attached and spreads between two spherical grains of radius R; it is a *pendular drop*. Surface tension exerts forces (represented by arrows) to pull at the deformed drop sticking to the grains. In reaction, forces on the grains are exerted; this tends to draw them together and increases the cohesion.

If a single layer of glass spheres is placed in a partially filled horizontal vessel (so that the water level lies below the sphere's diameter), the spheres will be attracted to each other by *capillarity** and form a compact pile.

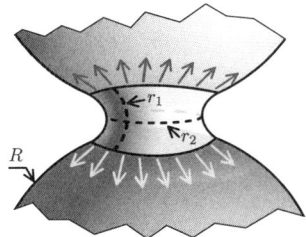

Figure 9.3
A meniscus of liquid water between two spherical grains. Its radius r_1 is small relative to r_2 and determines the negative sign of the Laplace-Young equation: the concave form of the water meniscus gives rise to a depression of the water; this attracts the spheres.

This attraction will disappear if the water level rises above the spheres' diameter as the menisci no longer exist.

It is possible to estimate the order of magnitude of the force working against grains' separation, which varies with the perimeter along which surface tension is exerted; this value is on the order of grains' radius R. Multiplying R by the water's surface tension γ_s yields the approximate force between two grains, $f_{cap} \sim \gamma_s R$. This force may be compared to the weight of a glass sphere hanging from another sphere of the same size and material that varies as R^3. These two forces have comparable values when the radius is of the order of one millimeter. Smaller spheres will hold by capillarity. That said, we should note that these spheres can easily slide against each other; even a slight layer of water will encourage a granular medium to reorganize.

To account for this result more precisely, scientists consider the curvature of the water bridge between grains in terms of the Laplace-Young equation. Here, in contrast to a drop that is simply convex, the sign of the average curvature $\mathcal{C} = 1/r_1 + 1/r_2$ is the same as that of the smallest radius r_1, which has a sign opposite to r_2 (cf. figure 9.3). Instead of the overpressure that occurs within a spherical grain, depression prevails in a capillary bridge and draws the two grains together.

For a given surface unit of a granular medium, the number of grains per unit surface increases, while their radius R diminishes as $1/R^2$. Accordingly, the total capillary force per surface unit will vary as $f_{cap}/R^2 = \gamma_s/R$, that is, as the inverse of the grains' radius: a sandcastle made with fine grains will exhibit greater strength than one made with coarse grains. Moisture content must also be considered: if water covers the grains entirely, liquid bridges disappear—and the effects of surface tension along with them. To return to our experiment with glass spheres on a horizontal surface: when water covers the grains, they detach and regain liberty. The cohesion between spheres is a function of the level below this limit. The equivalent of the water level for a wet granular material is the degree of saturation S, which is the ratio between the liquid volume and total volume of the pores (ranging between 0, for dry material, and 100%, for saturated material). What is the optimal water level to make a sandcastle hold?

EXPERIMENTAL PROOF

All along Copacabana beach, *criados*—the street children of Rio de Janeiro—will show sand sculptures for a few *reals* to the tourists walking along the beach; they last for a few days if kept moist. Serious research has been conducted on the optimal amount of water for ensuring the stability of structures like this. X-ray *tomography** uses a synchrotron light source to observe what is going within a pile—such recognition for the physics of sandcastles! These experiments account for the complexity of a mechanism that would seem trivial. At very low amounts of water, a so-called *hygroscopic* regime prevails: water attaches to grains chemically or through the effect of surface roughness. This regime isn't very important for sand. But when water rises to a certain percentage of the total pore volume, a *pendular* regime begins and capillary bridges form connections (figure 9.4). Bit by bit, a *funicular* regime sets in: menisci transform into clusters and form a continuous path across the granular medium. The number of bridges increases with the amount of water; at the same time, the force holding grains together decreases. Beyond a specific value, any increase of water and the number of bridges is compensated for by a decrease of capillary effects on grains, as the menisci become less curved. Hence, there is a plateau of cohesive force when the water content reaches approximately 10%. Anyone interested in sandcastle contests should also know that adding a little clay to the sand will improve cohesion. It can be further improved by adding a chemical agent to the water (for instance, sugar in solution); doing so makes the water more viscous and creates chemical interactions between molecules and active sites on grains—which delays the fatal loss of water that makes the overall structure collapse!

SIMULATED DROPS AND GRAINS

In addition to complex modes of observation and experiments like those performed with the synchrotron light sources, computers are another valuable resource for research on granular matter. Using computers, scientists can peer within the volume of the medium while varying the parameters of study, which can then be used directly in calculations. That said, such findings must still be confirmed experimentally. The back-and-forth relationship between empirical study and simulation progressively reveals the essential traits of physical phenomena, until a quantitative analysis can be obtained.

Can a computer help us predict how water penetrates within a granular medium? Indeed, it can—provided that we consider the motion of water in the space between grains. Theoretically, it should be possible to describe how each molecule of water moves. However, the grains that interest us here are much larger than atoms, and it would take a vast number of molecules before capillary bridges are formed numerically. Accordingly, scientists enlist a technique pioneered by Ludwig Boltzmann, the father of statistical physics and an ardent champion of the existence of atoms in the nineteenth century. Then one uses the probability density for finding molecules at a given point. If there are few molecules, it's a gas (vapor); if there are many molecules, it's a liquid. The Boltzmann equation governs the motion of these fluids. At a given position within the sample, the probability of finding molecules of water at any single moment in time will depend on their overall movement and, specifically, on the number of arrivals and departures at the point under examination. But because of collisions, certain molecules that should show up at this point are deflected beforehand—while others make it, even though they weren't in the molecule's initial trajectory. Figure 9.4 shows a 2D simulation of capillary bridges in a granular medium for different quantities of water using this so-called Lattice Boltzmann method.

a)　　　　　　　b)　　　　　　　c)

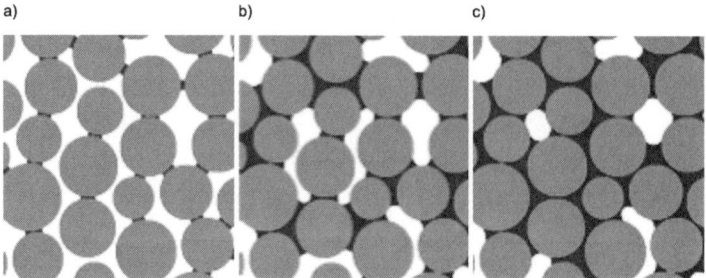

Figure 9.4

Three wetting states of a granular medium: a) a pendular regime formed by independent liquid bridges between pairs of grains; b) a funicular regime with liquid clusters isolated from each other; and c) a funicular regime in which the liquid is formed by "percolating" bridges—a continuous path of the water (in black) between one end of the medium and the other occurs.

If we consider the evolution of the number of bridges of a given size as a function of the degree of saturation, two regimes are found (figure 9.5a). In the pendular regime, the number of binary capillary bridges increases until a peak is reached. Indeed, capillary bridges initially form only at the contact points between grains but, as the amount of liquid is increased, new liquid bridges form between particles that are separated by a small gap. Above the peak value, binary bridges coalesce and yield bridges connecting three grains. This is the beginning of the funicular regime. At a higher level of saturation, the same process leads to bridges connecting four or more grains. Figure 9.5b shows these bridges in a two-dimensional simulation. When six grains are joined, stability starts to be compromised, and flooded grains (figure 9.6) appear; here the liquid no longer transmits any force.

Figure 9.5a shows a surprising result. Whatever the degree of saturation, bridges formed between two grains predominate, followed by bridges between three, then four ... If the sample continues to be saturated, a cluster of water greedier than others will continue to capture the surrounding water and become sufficiently large to traverse the system from one end to the other (figure 9.5b), as in the case of a percolation threshold, discussed in chapter 6. Finally, at high levels of saturation, a moment comes when there is no longer enough air to make bridges; now, only bubbles trapped between grains, maintaining local suction, remain.

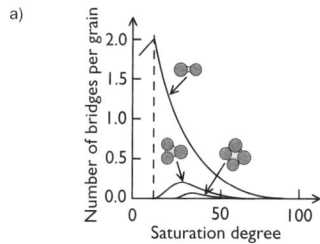

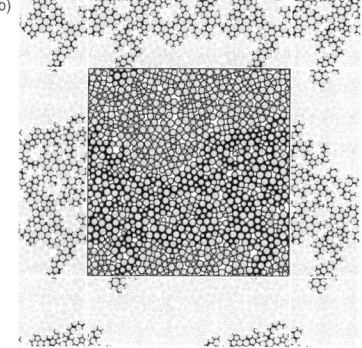

Figure 9.5

a) Evolution of the number of capillary bridges connected to two, three, and four grains divided by the total number of grains as a function of the saturation degree. b) The state of water (in black) at the moment when the largest water cluster has crossed the sample entirely. The sample has been replicated periodically (the edge of replicated areas is rendered in light gray).

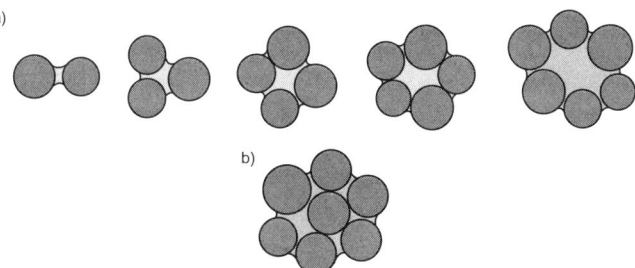

Figure 9.6

a) Examples of liquid bridges between grains. b) Liquid bridge incorporating a flooded grain.

MACROSCOPIC COHESION

At some point, you may have tried making a *croquembouche*, a delicate French pastry. This work of culinary artistry consists of a extremely steep and airy pyramid of cream puffs—"grains" all the same size—which only holds up if caramelized sugar is added between points of contact to ensure the structure's cohesion. In a wetted soil, water is what makes the grains stick together, thanks to the force of capillary attraction f_{cap}.

In chapter 5 (figure 5.9), we saw that in the absence of cohesion between grains, the shear stress τ in steady flow of a granular material is given by $\tau = \mu_c P$, where P is the confining pressure and μ_c is the coefficient of friction. When "sticking" occurs at the level of contacts, it's necessary to exert a sufficiently large shear stress in order to overcome not only the friction but also the effect of cohesion, which is measured by a parameter c, known as *Coulomb cohesion*. Here, the preceding equation takes on a more general form: $\tau = \mu_c P + c$. In figure 9.7, it is represented by the full line. This relation means that the shear stress is simply equal to the Coulomb cohesion c when the confining pressure vanishes. The equation may also be written as $\tau / P = \mu_c + c / P$. In a heap, the relative cohesion c / P contributes to raising the slope angle above ϕ_c, if the sand is dry; wet sand can remain in a state of equilibrium when slopes are extremely steep, even vertical.

Hence, the effect of cohesion on the maximum slope angle depends not only on the cohesive stress c but also on the weight through the ratio c / P, so that increasing P tends to reduce cohesion effects. Capillary cohesion is enough for building sandcastles on the beach. However,

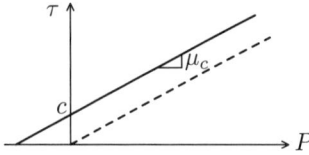

Figure 9.7
The Mohr-Coulomb criterion: shear stress τ varies linearly with normal pressure P applied from outside or due to the weight of the grains. The dotted line represents the case of a dry granular medium for which cohesion c is zero. The full line corresponds to the case where cohesion c accounts for an adhesive effect and shear strength even when P is negligible.

the same cohesion will be negligible for large piles of grains in which the vertical stress due to weight plays the dominant role. At the seaside, the pressure at the base of a sandcastle increases gradually as wet sand is added. Beyond a certain height, the ratio c/P becomes too small, and the structure collapses. For grains 0.1 mm in diameter, capillary cohesion is enough to make a column with a maximum height of 30 cm. This, incidentally, is the size of the little pails sold for making sandcastles! To go beyond this limit, finer sand is required since, as we have already seen, cohesion varies as the inverse of grain size. Capillary cohesion reaches its highest values in clays where the aggregates of platelets are nanometric in size (see figure 9.14).

Cohesion influences the packing fraction and dilatancy of granular media. In the presence of capillary forces, single grains can find a position of equilibrium with fewer contacts. On this basis, one is able to obtain configurations that are looser and less compact than in the case of dry grains. The material's capacity to densify by compression is essential for manufacturing items such as salt pellets for water softeners or pharmaceutical tablets.

WATER IN THE SOIL

WETTED SOIL

When one plunges one end of a glass tube into a liquid that wets it, the smaller the tube's diameter is the higher water will rise in the tube. This demonstrates a phenomenon known as *capillary suction* or *imbibition*.* The force leading to displacement of the liquid layer is the Laplace force, which is manifested by the concave shape of the meniscus at the top of the tube. This same phenomenon may be observed by dipping the edge of a sugar cube into coffee. It also plays an important role in making water available to plants in soil.

The maximum quantity of water that a soil can retain or store after the overflow has infiltrated (or "drained") into ground water is called *field capacity* (or *retention capacity*). This parameter is important for agriculture: together with permeability (to be discussed in chapter 11), these parameters are used for estimating irrigation levels. The retention curve permits us to describe, in greater detail, the way the material retains or

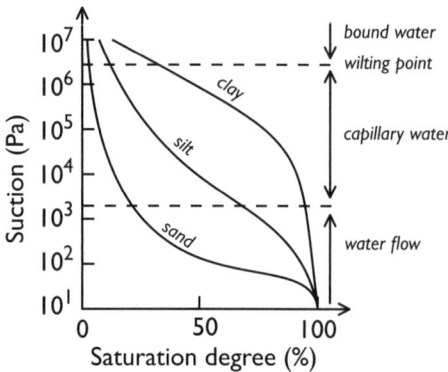

Figure 9.8

Evolution of the degree of saturation in water as a function of capillary pressure, or suction, for soils ranging from coarse (sand) to fine (clay).

releases the water contained in the pore space. This curve, which relates the degree of saturation to *suction*, depends on the size of grains and, especially, on the presence of clay, the finest particles found in soil. Likewise, the lower the quantity of water is, the greater the effort required to aspirate it. Plants that draw water from the soil do so at a cost: the roots need to overcome capillary forces attaching it to solids. If the amount of water is too low, it becomes impossible to extract it, and the plant withers: the soil is said to be at the *wilting point*. In figure 9.9, we can see that, as a function of the moisture level and grain size of the soil, water may be present in the form of bound water, which is strongly connected to the soil, in the form of capillary water, which is readily available to plants, or in the form of gravity-fed water, that is, water that grains do not retain.

MECHANICAL STATES OF WET SOIL

When you walk across the beach toward the water, at first the sand is dry and deforms wherever your foot lands—which makes progress arduous. Gradually, the sand grows moister, and the beach's surface becomes almost flat; walking proves easier and leaves visible traces. Finally, right at the edge of the sea, the sand seems to fluidify: water saturates the spaces between grains, and steps leave holes that the sand and water will quickly fill again.

We can distinguish between at least three mechanical behaviors that depend on the nature of grains and relative water quantity. If the amount of water is sufficient, the mixture will flow between one's fingers like pancake batter. In contrast, a weak amount of water yields a solid that will break if deformed. Between these two extreme phases, an intermediate state of plasticity exists that will yield a paste if kneaded. Transitions between solid and plastic, which are known as *Atterberg Limits*, form the basis for describing wet granular media. These limit values depend, in particular, on the speed at which water migrates within the medium and the physico-chemical nature of the grains. The presence of clay—fine particles of organic matter—plays a key role in the plastic phase.

COMPACTING WET GRAINS

An optimal quantity of water exists, at which the packing fraction of the grain assemblies reaches a maximum value as a result of compaction. In 1929, Ralph Proctor, a field engineer in California, set forth a practical means of analyzing compaction for civil engineers (the California impact test), who must pile soils where buildings and roads are to be made. This simple test involves taking a soil sample, with a given water content, placing it in a mold and subjecting it to the repeated impact of a calibrated mass. Energy is monitored in terms of the precise number of blows applied. In a second step, the soil's dry density is assessed. By repeating the operation, one can determine the evolution of dry density for different values of water content—in other words, the weight of water with respect to the weight of grains. The resulting curve allows one to find the water content that will enable the best possible compaction. This maximum volume fraction is called the *Proctor optimum* (figure 9.9).

The preceding description is quite general. Other compaction procedures yield similar curves, but they depend on the composition of granular material and energy applied. Recent studies have shown the utility of generalizing such analysis to all procedures involving mixtures of grains in the presence of a liquid. The resulting diagram, which describes the phases of a wet granular medium, is called *hydrotextural*. It enables one to follow the formation of aggregates of couscous, the manufacture of pharmaceutical tablets, or any other procedure for lending a material a texture

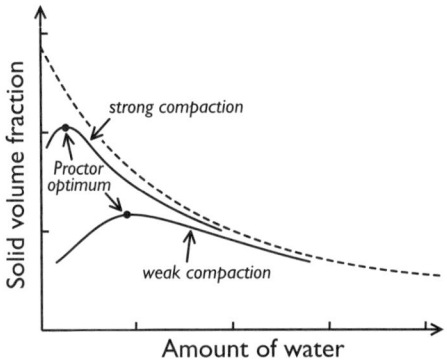

Figure 9.9
The Proctor test for determining the dependence of solid volume fraction on water content for two compaction energies. The dotted, asymptotic curve represents a limit case where water would saturate the material.

by controlling the ratio between water and powder and the energy it takes for kneading, blending, and compounding. Figure 9.10 provides an example of the experimental results of such a process. Take polenta, semolina, or any other powder that is relatively insoluble and knead it with a little water. At low levels of moisture content, the medium's packing fraction will initially decrease—sometimes yielding a spectacular expansion of volume. A similar phenomenon, well known in mining and excavation, is called *sand proliferation*: it's not uncommon to find oneself with 30% more soil than the volume of the hole. This is a manifestation of the work performed by capillary forces, which prevent earth deformed by digging from redensifying. If water is added continuously, proliferation decreases and the material densifies until it reaches an optimum equivalent to the Proctor effect shown in figure 9.9. Finally, a regime sets in where the quantity of water occupies all the space between grains, which gradually become dispersed in the liquid.

As we have seen, the packing fraction of a dry granular medium depends on the method of filling employed. The same holds for a wet medium displaying capillary effects; here, a rich array of states exists. If one starts with an ensemble of grains that are wetted by adding water—by pulverization, for example—mechanisms of aggregation will set in, especially through local capillary forces, and change the positions of grains

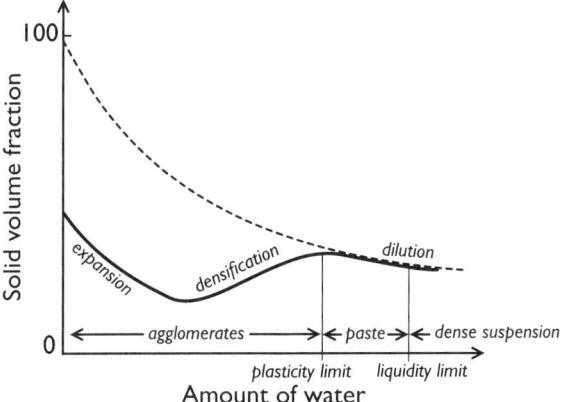

Figure 9.10

Phase diagram of a wet granular packing representing the variation of solid volume fraction as a function of water content. For an amount of water below the plasticity limit, agglomerates form. For higher amounts, the agglomerates coalesce and form a paste. As even larger amounts of water are added, the solid fraction further declines with increasing dilution of the suspension.

and their ultimate packing fraction. This is why mixing and kneading wet granular media is a delicate operation. A baker making dough knows all about it: the end product depends on action applied to homogenize the distribution of water and create the network of gluten that gives bread its elasticity.

EARTH STRUCTURES

One important aim of this chapter—beyond seaside amusements and demonstrating the effects of capillarity in granular media—is to explore ways of stabilizing inclined soils against erosion, building embankments that hold up, and consolidating structures made from raw earth. This last application is absolutely essential inasmuch as it bears on the lives of some two billion human beings. In each instance, effort is made to increase intergranular cohesion, which, as we have noted, depends on grain size and water content. As we will see a bit later, there is also a host of effective methods for improving material resistance by adding foreign elements that consolidate grain compounds.

ARMING THE EARTH

But we shouldn't forget games at the beach just yet. Henri Vidal, an architectural engineer, had a flash of inspiration while playing with his children: he recognized that the sand piles they were building could be strengthened by adding pine needles. Eventually, this led him to register a patent for *reinforced earth*! Subsequent developments have produced far-ranging results, especially in civil engineering. Major improvements concern roads, which can deform and develop ruts as a result of traffic after heavy rainfall, and steep highway embankments, which are subject to landslides. Metal framework reinforcing the cohesion of granular media in walls started being used in the 1970s. Geotextiles placed at different soil levels have helped to reduce the effects of corrugation that occur when wheels tear at roadways. *Texsol*, another crucial innovation, is a homogeneous mixture of synthetic threads and sand that enables much steeper slopes than sand alone. Plants can also be grown to encourage a durable root system and enhance stability.

BUILDING WITH EARTH

The title of this section is inspired by an insightful and beautiful book written by two young researchers at the Grenoble School of Architecture. Their ambitious project was devoted to finding ways to improve structures made from earth, which people have been making for ages. If, as we just saw, an engineer deserves credit for innovations that benefit modern industry, the idea of adding fibers to soil to promote cohesion is ancient. (The Great Wall of China, built and rebuilt over the course of more than twenty centuries, provides a spectacular example; since it occupies dry regions, it manages to do without using capillarity.) The Grenoble team managed to erect tall columns of dry sand by inserting sheets of cardboard at regular intervals (figure 9.11). On the same principle, skyscrapers in Yemen and mosques in West Africa display imposing architecture, even if they require regular maintenance (figure 9.12). This traditional mode of building, likely practiced since the Neolithic, still houses billions of human beings. In the Grenoble region, houses made from earth that is packed but not reinforced—called "pisé de terre," or *rammed earth*—dot the countryside. The point is not to replace this native

Figure 9.11
The stability of this structure made from dry sand, some dozen meters high, is due to the presence of fine layers of paper placed regularly between the layers of sand and the struts, which prevent bowing.

Figure 9.12
The Great Mosque of Djenne in Mali, constructed from bricks of raw earth dried in the sun (adobe).

form of local architecture, which draws on resources provided by nature, but to improve cohesion as well as resistance to both rain above and soil moisture below.

The subject is of great importance. It is reasonable to imagine that new modes of construction will be found that use locally available materials and respect the environment: the operation occurs at room temperature, as it does not make use of cement (production of which represents a major source of CO_2). Some contemporary architects have understood the value of building with earth and have used it in designing ambitious housing projects and monuments.

REINFORCED CONSTRUCTIONS

Walls made from bricks of raw earth go back some ten thousand years in Asia. Significantly, they include plant stalks before bricks are stacked and dried. More recent developments have made it possible to manufacture bricks with better mechanical and thermal qualities. This is the case, for instance, for bricks that combine local soil and typha (an invasive plant growing on the banks of the Senegal River). These bricks are strong enough to withstand repeated hammer blows; this durability derives from the way the fibers are treated beforehand, to stick to the wet earth. The project is win–win because it manages to use a noxious material to beneficial ends—both for people and the environment. Adding other waste material only improves the quality of concrete. An essential ingredient that will receive separate treatment is the clay present in brickmaking; bricks of raw earth combining vegetal refuse and clay, then dried in the sun, are known as *adobe*.

With names that vary by region and tradition, building practices employ different strategies to combine earth, clay, and various admixtures. *Cob* uses a framework of fibers or flexible wood in a trellis to make walls. *Timber framing* consists of a visible (and decorative) array of intersecting beams, with cob filling the spaces in between (the French term for the same combination of materials is *bauge*). In each case, the presence of clay contributes to the cohesion of the whole.

GRAINS TAKE ROOT

Nature can do what also occurs by human artifice: roots stabilize the soil and, at the same time, absorb rainwater, which limits erosion. *Beach grass*, for example, is planted to stabilize dunes with vertical roots, while a system of thick rhizomes covers the surface and horizontal shoots, *stolons*, extend underneath. We will not explore the rich variety of such systems here, but they play a key role in the way roots interact with soils and subsoils.

Exchange between roots—which can amount to thousands of meters under just one square meter of soil—is essential for the life and health of plants. It can extend to tens of meters within the soil in desertic zones. This "dialogue" takes the form of a rhizosphere, in which microorganisms, water, and minerals influence the aerial growth with which we are more familiar. Whether considering the plant above or the unsaturated soil where its roots are plunged, research on transfers of mass, energy, and soil mechanics will benefit from the approaches we have been discussing.

The variety of situations takes us back to the model of glass spheres, now coupled with a nutrient medium of the kind used for off-ground cultivation (figure 9.13). Here, the grains grow by the roots and stalk; one can follow their "subterranean" growth in a transparent container. The root apex grows by creating new cells and elongating old ones. If the glass

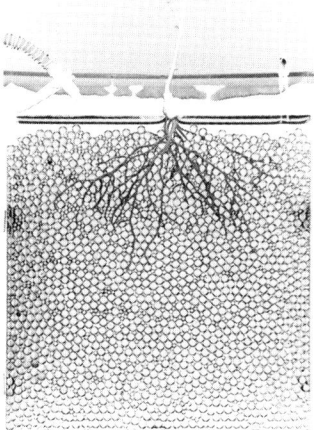

Figure 9.13
Chickpea roots growing in a thin layer of grains. The roots work their way between grains and displace or compress them to become anchored in the soil.

spheres are replaced by photoelastic grains, one can see the effect of radial and axial stresses that roots exert on grains as they develop. In the figure, the root is developing toward the lower part of the vessel, which holds a nutrient solution; in the process, it gives rise to stresses on discs, which symbolize mechanical obstacles that slow, or even prevent, further growth.

A further parameter is geometrical cohesion: as roots twist and turn, they stabilize soil components, somewhat in the manner of nonconvex, elongated grains. In this symbiotic relationship, plant and soil stand together to create better mechanical strength.

INCREASING COMPACTNESS

Water plays an ambivalent role in earthen structures: it is responsible for the solidity of a pile of grains … provided it's used in moderation! One way to decrease the water content of a pile of grains is to *reduce* porosity, which means leaving as few voids between elements as possible. As our model with spheres (chapter 3) has shown, packing fraction remains roughly constant (between 60 and 64%) if elements with the same diameter are used; if we mix spheres of different diameters, it becomes larger. In light of this fact, fillings on multiple scales are employed. Apollonian packing provides a theoretical model that would entail zero porosity, if it were possible to reduce such packings down to infinitely fine grains (figure 3.14). When building structures from earth, effort is made to achieve the broadest possible distribution of sizes. The next chapter will explore how a seemingly trivial decision to add extremely fine silica powder (industrial waste) to a mixture of grains of various sizes has played a revolutionary role in developing high-performance concrete.

CLAY

It's fine to reinforce a structure with fibers and reduce porosity with a mixture of grains of different sizes, but it's crucial to not forget the essential ingredient in this partnership—to make a solid structure out of earth with particles of soil that have highly variable sizes, we need the smallest ones: clay.

Silicon and aluminum are omnipresent in the earth's crust (geologists refer to them as *SiAl*). Combined with other elements, especially metals, they form chemical edifices that consist of platelets of clay. The latter provides the smallest-scale component of sedimentary rocks and soils, with grain sizes of just a few micrometers. In foundational myths, clay (*Adam* derives from *adamah*, "earth" or "soil") brings forth human beings. It has always provided a resource used in ceramics and other types of construction for achieving solidity and durability.

The Chemistry of Clay

Mineralogical chemistry explains how clay is structured like Lego pieces, on the basis of planes combining tetrahedra and octahedra; they form around positive ions such as silica and aluminum surrounded by negative ions of hydroxide, oxygen, and water. A sheet represents a pile, some dozen nanometers thick, made of two or three chemical layers. Stacks of sheets make grains a few microns in extension, resembling a thin book made from Bible paper (also known as scritta paper). Moving beyond this schematic description, a wide array of structures exists, depending on the number of elementary layers in a sheet and whether the charges are at equilibrium—as is the case for kaolin, a fine, soft white, clay—or enlist ions from outside to ensure electrical neutrality. Clays of the latter kind are able to inflate or deflate depending on ambient physico-chemical conditions. Such adsorption proves catastrophic in the case of sloping terrain, where the clayey soil fluidizes and causes landslides. In contrast, during a dry season, the deficit of absorbed water takes the form of cracks that result from the clay's retraction.

In this context, we can call on a specialist of material in earth and mud, Henri Van Damme, who stressed the function of water in contributing to the cohesion of classic granular media. The smaller the grain scale, the more important capillary effects prove to be. Water lodges in the spaces between sheets of clay; inasmuch as separations between layers are weaker than between grains, capillary attraction is stronger. Figure 9.14 shows how capillary effects manifest themselves, ensuring much stronger cohesion than in the case of a sandcastle. Moreover, clay holds water, even in a dry environment: in the desert, where water content is minimal, it's still there in condensed form, in channels a few nanometers wide. On this scale, clay remains permanently wet and cohesion is ensured. The contents of a single can of a soft drink are enough to wet a cubic meter of earth! Cohesion is strengthened by electrical forces that derive from

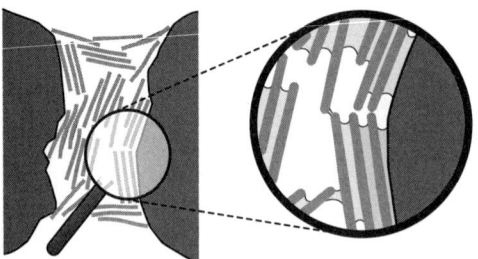

Figure 9.14
Clayey microstructure displaying capillary effects at the level of sheets ensuring cohesion between grains.

components of clay in ionic form. This feature entails a whole array of new behaviors. For instance, by varying the pH of the medium, one can modify the rheological behavior of an entire clay suspension—from a compact mud to a concentrated solution of grains. It is the triumph of capillarity at small scales!

In this chapter, we have encountered the effects of low cohesion between materials in the sense that, despite the presence of cohesion, the grains can still slide over each other and separate if the medium is subjected to enough stress. Chapter 10 will address the effects of high cohesion. The difference is slight, and continuity exists between these limits. The "kinetic sand" sold in toy stores offers an example: a very thin layer of silicone oil covering silica grains ensures cohesion and takes the place of water's capillarity: there's no need to go to the beach to build such sculptures in the absence of water! The ease with which this material can be deformed is an example of plasticity—like the clay potters spin on their wheels.

10

STICKY GRAINS

Then the earth, compressed by the air so water cannot dissolve it, forms stones—the loveliest being transparent stone consisting of equal and regular parts, and the ugliest having the opposite qualities. The kind stripped of all its moisture by a rapid fire, a formation more brittle than the other, is called brick.

Plato, *Timaeus*

Nougat, granola bars, oatmeala cookies. … Food offers countless examples of grains being mixed with various binding agents to make cohesive textures agreeable to the palate. The construction and pharmaceutical industries also manipulate "sticky grains" on a daily basis. There are many ways to obtain materials by varying the proportions of three phases: the grains themselves, a matrix ensuring their cohesion, and voids filled with air. Figure 10.1 provides a few assemblies of this kind.

A packing of grains, which we have been speaking of at some length now, represents the first stage in the natural or artificial creation of materials in which permanent contacts—bonds—are established between particles. This chapter examines such materials obtained by consolidation.

One method for binding grains together involves filling the porous medium with a sufficiently fluid polymeric resin and then draining off as much liquid as possible. Some of the liquid will remain stuck at the level of contacts between grains by capillarity (cf. the pendular drops discussed in chapter 9), while leaving voids in larger spaces. After polymerization,

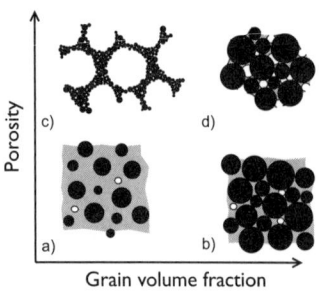

Figure 10.1

Examples of microstructures obtained with different quantities of matrix (gray), grains (black), and voids (white): a) composite made from a matrix, diluted grains, and voids; b) cemented granular matter in which grains in the matrix are in contact; c) granular foam formed by a skeleton of cemented grains surrounding large pores, and d) cohesive granular material with mechanical strength deriving from localized adhesion at the level of contacts.

this adhesive attaches grains permanently, with spaces preserved in between. The granular medium has been consolidated. This type of bonding may be observed in *sedimentary rock,** which will be discussed further on: water charged with mineral salts circulating between particles gradually leaves solid deposits that join grains together, sometimes by inducing a chemical reaction at the level of contacts.

In the manufacture of concrete, the granular medium composed of sand and grains of various sizes is mixed with a reactive cement powder in the presence of water: when the compound reacts with the water, it forms crystals that connect grains. When the reaction ends, however, there is a residual porosity that should be kept as low as possible. The remaining water no longer plays a role; it disappears, in part, in the course of drying. Mechanical tests of a concrete structure are conducted only after waiting a few weeks. A final process is sintering, in which no outside binding material is used.

SINTERED MATERIALS

THE SINTERING PROCESS

Without being conscious of it, when you take a handful of fresh snow to make a ball out of it, sintering is at work. Liquid water between grains that turns into ice—or manual pressure putting grains in contact and joining them—ensures the further coherence of the projectile. A trivial example? Not at all. In the nineteenth century, it provided the object of a scientific debate between Michael Faraday and Lord Kelvin, who disagreed about the process of consolidation. Employed in the manufacture of a host of materials under controlled conditions, sintering transforms a packing of compressed particles of powder into a consistent whole by raising either temperature or pressure. The sintering temperature is lower than the grains' bulk melting temperature. Materials obtained by means of joining grains to each other permanently are called *sintered*.

Sintering produces a bond between grains that were initially free. The displacement of matter necessary to achieve this result can occur in the particle volume; more commonly, however, it takes place on the surface, where atoms are freer (figure 10.2a). Surface mobility is influenced in large part by the effects of surface tension (see the discussion of Laplace-Young's law in chapter 9), which expresses the effects of pressure resulting from the curvature of interfaces between the solid and voids. Pronounced curvature effects occur when a material is sintered: the space between grains forms a concave mensiscus contrasting with the convexity of the grains themselves (figure 10.2b). Atoms migrate from the convex zone to the concave zone, where attachment is better. At the same time, the "necks" that form between particles thicken.

These surface atoms can also travel "by air" (figure 10.2). Because of the curvature effect, vapor of the solid's atoms exerts greater pressure on a

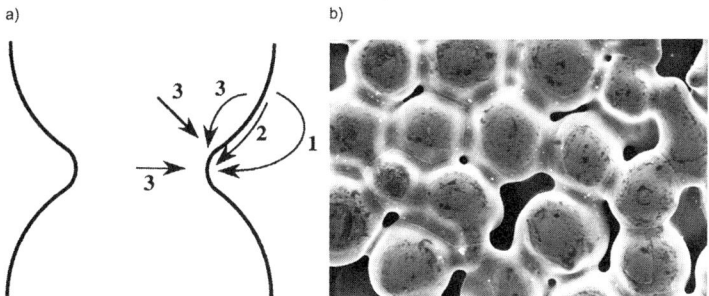

a)   b)

Figure 10.2
a) Different transfer modes of matter toward a sintering neck in the course of formation: 1) vapor transport; 2) surface diffusion; 3) volume diffusion. b) Microscope view of sintering necks between grains of copper.

convex surface than on a concave one; on the former, the atoms are more exposed. Accordingly, atoms move into the porous space, which also thickens the neck between particles and consolidates the packing. The photo in figure 10.2b shows these necks at an early stage of the process.

Subsequently, the necks created by sintering thicken. Pores close, and porosity decreases. Complete densification marks the final stage, when all porosity has been eliminated: the result is a compact solid with defects resulting from the process. Needless to say, the question arises of how to eliminate vapor in closed pores that are surrounded by solid matter: applying extremely high pressure can make residual gas escape through interconnected voids.

CHARACTERISTICS OF SINTERED MATERIAL

It is possible to go beyond a simple microscopic observation. A quantitative analysis of cross sections enables certain volume characteristics of the material to be determined. This is the object of *stereology*,* a branch of science that received its patent of nobility more than two hundred years ago in the works of Leonhard Euler and Gaspard Monge. The operation consists of using two-dimensional cuts of a random composite material to obtain a three-dimensional description. The next section will examine how such applications can be used for *sintered materials*.*

In addition to this classical method for determining the average bulk properties of a composite, recent developments in tomography (also used in medicine—MRI and X-ray tomography, for instance) have been used by specialists of material science (figure 10.3). This approach involves an array of images taken from different angles; by means of series of computerized images, it is possible to arrive at the object's structure in three dimensions. The process is known as solving an *inverse problem*: 2D images of a sample are used to go back to its bulk characteristics if certain properties, such as isotropy of the material, are met. These characteristics are described in the following.

Packing Fraction and Specific Surface Area Two average parameters are essential for characterizing the geometry of a porous medium.

• The *volume fractions* of phases—solid and void—are the packing fraction of the medium or its complement to unity, its porosity. For a

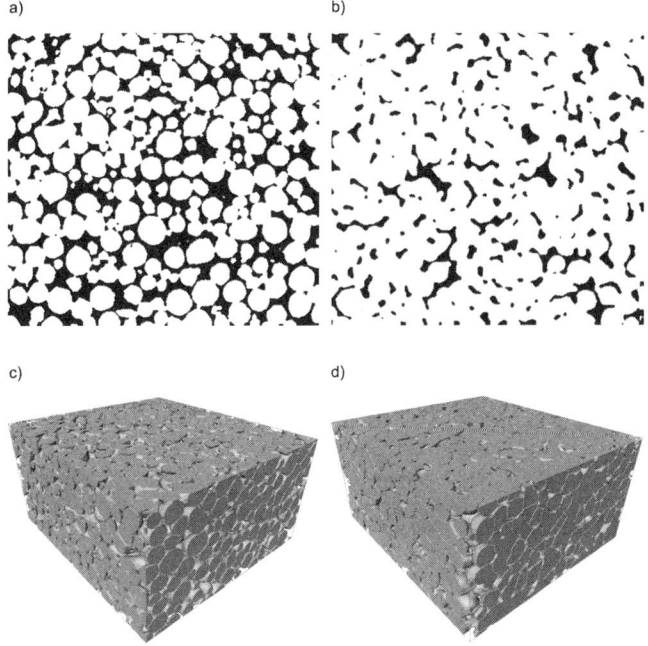

Figure 10.3
Sections of an assemblage of copper beads a) before and b) after sintering. c) and d) are corresponding tomographic reconstructions in three dimensions.

pile of identical spheres, the packing fraction of the medium increases from 60% to 100% from the beginning of the sintering process to the final, fully densified material.

- *Specific surface area** is the area per unit of volume of the interface zone between these two phases. For a packing of spheres in contact, it is proportional to the inverse of the spheres' radius. Thus, for one-micron grains, it is of the order of 1,000 square meters per cubic decimeter. This characteristic is involved when the surface properties of granular media are at play, for example, in adsorption of a gas (for activated carbon or zeolites).

Curvature and Genus Two other geometrical parameters are necessary for characterizing a sintered material. The *mean curvature* of the sample as a whole is defined by analogy with the Laplace-Young law (see chapter 9). For weak levels of sintering, the grains are practically isolated from each other, and thus a large positive mean curvature is obtained. As sintering proceeds, concave zones appear where necks form and leave voids. These holes close up and disappear, ultimately yielding a compact object. In this way, curvature passes from a large positive value to zero, where curvature effects in both directions compensate for each other, then to a negative value reflecting the holes that remain; finally, it vanishes when these holes are reabsorbed.

A fourth parameter, *genus*, concerns the *topological* characteristic of the medium; it allows us to follow how particles connect to each other in the sintering process. For example, in an ensemble of four grains in contact forming a tetrahedron, the genus is three: three bonds between grains can be made while keeping the structure in a single piece. In a given grain assembly, genus concerns the maximum number of bonds that may be suppressed without fully disconnecting the structure.

These four parameters can be evaluated by means of random sections of a sintered material, as shown in figure 10.3.

PREPARING SINTERED MATERIAL

Let's return to how sintered materials are made—whether sintering occurs as the evolution of granular media themselves under the effect of temperature and pressure, or as a result of adding an external material—a glue—in order to join grains. Historically, diverse modes of preparation have been employed on a broad range of materials. As far back as 2,500 years ago, our ancestors were already making jewels from powders of refractory materials, which they didn't need to heat to reach the melting temperature of the bulk. They would place grains in molds that represented the desired form in negative and then fire them at high temperature. A tall iron column in Delhi erected some three hundred years before the Common Era—which is still standing today—required six tons of iron powder.

When different metals are combined, they can form phases such as alloys—or not. If the melting points are different enough, as in the case of copper and tin, sintering takes place when only one of the phases (tin) is in a molten state. By this means, porosity is eliminated: in the course of the operation, a double capillary effect makes the solid grains (copper) tend to bond while, simultaneously, the liquid (tin) creeps into vacant pores.

The end result will prove satisfactory only if grains have been properly arranged. Filling should be accompanied by shaking. Sometimes it's necessary to apply pressure, as well. *Uniaxial compression* may be applied between the upper and lower extremities of a vertical cylindrical mold, or *hydrostatic compression,** which involves placing the powder in a flexible mold and subjecting the outer casing as a whole to an overpressure of liquid or gas. The level can reach up to 10,000 times atmospheric pressure; an additive helps to reduce friction and obtain the same compaction at lower pressures. The resulting material, which is hard but brittle, is subjected to heat treatment at this stage or later. We should also mention one other procedure—*slip casting*—in ceramics: here, powder is placed in a liquid, and the physico-chemical conditions (pH value, additives) are adjusted so the pasty mixture will remain fluid. In turn, the mixture is poured into a porous mold, which absorbs the solvent. The dried material is removed and fired—leaving the mold ready for the next use.

While still employed in traditional metallurgical processes, sintering now provides the basis for manufacturing both high-tech materials and everyday ones such as medical substitutes for bones or teeth, ceramic knives, and pharmaceutical tablets. Accordingly, specialists believe that granular solids will represent one of the major sectors of industry in tomorrow's material technology.

POROUS SINTERED MATERIALS

To achieve high porosity, a low sintering level is required. Permeability, to be discussed in chapter 11, largely depends on porosity. As we have seen, grains of matter may be joined with very little adhesive. If grains in a wet medium are in contact, the quantity of water does not affect cohesion much. In consequence, it is possible to obtain sintered materials with a high percentage of voids between grains that still show high mechanical strength.

In a copper heat exchanger, high permeability is sought, as well as a large contact surface between the fluid exchanging heat with porous material. One application is the bushings used in the contacts between moving parts like axles: a sintered bronze bearing with a porosity over 20% is permanently filled with a viscous oil that serves as a lubricant.

The dynamic friction in the course of rotations heats the bearing, and the oil becomes more fluid, forming a film on the surface of the metal axis that reduces friction. When this activity stops, the oil reintegrates the pores after cooling down. Another example is the manufacture of sintered titanium prostheses. When the amount of void is large enough, the elastic modulus approaches that of human bone; in this way, interactions with the living tissues that come to colonize the pores are encouraged, and stresses are distributed more evenly between the prosthesis and the biological body.

In contrast to unconsolidated granular packings, the contacts within sintered objects are established once and for all. More specifically, the number of contacts between grains does not depend on pressure; local relative rotations of particles are limited by elastic restoring forces, which involve the material between the grains. The elasticity of sintered material resembles that of a metal grid with bars cut out at random—which may be simulated with a model of vector percolation (cf. chapter 7).

CERAMICS

Pottery, which is one of the oldest testimonies to human endeavor, may be considered a form of sintering. The terrain is so vast we must content ourselves with just a few remarks. Undoubtedly, the first artisans got the idea from observing soil rich in clay that solidified in the heat of a fireplace. The word *ceramics* comes from the Greek word *keramos*, which means "potter's earth" or "clay." Clay is the key element of soils sprung from inorganic chemical processes, combined with minerals. Its consistency before firing—the critical part of what is called *soft matter**—depends on composition and water content. To make a consistent paste, artisans today do the same as their forebears. Chapter 9 examined the various behaviors obtained by combinations of earth and water kneaded by hand. Potters seek a plastic phase that they will be able to work and give form—like playdough. The ultimate quality of products will depend on the water content and clay composition, as well as the firing temperature. It's also necessary to eliminate residual pockets of air, which can cause damage in the course of firing. In order to ensure cohesion and prevent the clay from cracking, adding grains (such as sand, shells, refuse) helps sintering and improves cohesion.

Terracotta, which has been known since antiquity, is made in furnaces with a heat below 1,000 degrees. Terracotta itself remains permeable; to remedy this feature, artisans long ago developed the technique of glazing, which involves firing objects a second time with a continuous suspension of pigments on their surface. As temperature rises, it fuses with the clay and produces more resistant objects. Transparent ceramics (porcelain) were developed in China around two thousand years ago, and were not copied and manufactured in France until the eighteenth century. Porcelain consists of kaolin mixed with quartz and feldspar; the latter, in particular, lowers the melting point. Many other kinds of ceramics have been invented since then. Today, ceramics resulting from sintering refractory materials play a vital role in industry, especially when resistance to high temperature is required.

3D PRINTING BY SINTERING

Until recently, sintering occurred in molds or matrixes filled with powder beforehand. But there are also more modern, "additive" processes, which make it possible to fabricate three-dimensional objects starting from powders, stalks, or fibers. Unlike traditional methods of machining, which extract matter and generate waste in the form of chips or shavings, these new procedures construct the object by the progressive addition of matter. The principle of 3D printing is to glue on successive layers of grains with the help of a localized source of energy. Laser sintering, for instance, occurs in a heated enclosure containing metal powder spread on a plateau. An infrared laser beam raises the temperature locally and melts the surface of the powder, which rigidly attaches the grains to each other by capillarity and solidifies when they cool. The same operation is repeated layer by layer, until the piece has been completed. In addition to the simplicity with which complex mechanical parts are obtained, methods of this sort have the advantage of being able to produce pieces made from several powders. To this end, the object is fashioned in successive layers, with varying amounts of constituent elements, yielding "alloys" with a continuous transition between components. The process proves especially interesting when the task is to improve thermodynamic properties of, or limit interfaces between, metals that are difficult to join because they cannot be welded, or because their dilatancy coefficients are too different and subject to fluctuations of temperature.

Markus Kayser, at MIT, has devised a similar technique for a 3D printer called the Solar Sinter, which, as the name indicates, works with solar energy. This project has managed to produce objects from desert sand cemented only by resources on site. The long-term prospects for processes of this kind are quite interesting—say, on planets without liquid water ...

CONCRETE

Concrete is a very important example of a medium consolidated by adding outside material to ensure granular cohesion.

STICKING GRAINS TOGETHER

Plaster, lime, and cement are classic ingredients for construction used as *bonding agents* between solid grains of different sizes. In the presence of water, the mineral salts that constitute these phases form solid crystalline hydrates that act as adhesives. We will limit discussion to the case of concretes obtained by mixing nonreactive granulates and bonding agents: classical *hydraulic binders* involve reaction with water, but the process can also be performed under conditions of immersion (for instance, under the pylon of a bridge). In fact, the word *concrete* has a broad range of meaning: we speak of *asphalt concrete* (or *bituminous concrete*) when a hydrocarbon binder is used, and of *Earth concrete* where clay acts on the consolidation of solid grains at room temperature (cf. chapter 9).

A LONG HISTORY, INTERRUPTED

The invention of concretes isn't recent, but it has known ups and downs. The Pantheon of Rome, with its cupola eighteen meters in diameter, has stood the test of time. The recipe was a mixture of lime and silicates, together with especially lightweight pozzolan sand from neighboring volcanos. The architect Vitruvius was a reknowned expert during the Roman age. His epoch-making treatise *De architectura*, dedicated to the emperor Augustus, gave rise to projects for centuries to come, including the vault of Hagia Sophia in Constantinople (modern-day Istanbul). But later, the use of concrete was forgotten; wood and stones took the place of reactive mixtures of minerals for constructing large-scale edifices. It was not until the nineteenth century—with the pioneering preparation of artificial cement by Louis Vicat, and later by Le Châtelier's analysis of chemical reactions that occur during bond formation between grains—that concrete returned in full vigor with mixtures of lime, silica, and alumina (which exist abundantly as raw materials, but must be heat-treated before use).

NEW HIGH-PERFORMANCE CONCRETES
Inventing better concretes is the aim of research by specialists in granular materials and civil engineering, an endeavor with considerable practical implications. In fact, it is more accurate to speak of concretes than of concrete. Classic concretes and mortars contain grains of different sizes: sand, gravel, and ballast, for example. Concrete can also incorporate residues, for instance shards of pottery and powders—provided there is no risk of pollution through leaching (or lixiviation) from the circulation of water. A good concrete will contain fine, medium-sized, and coarse grains simultaneously (figure 10.4). Overly homogeneous mixtures of very fine sands—the particles in dunes, for instance—lead to poor resistance. The goal is to come as close as possible to Apollonian packing, discussed in chapter 3; porosity would fall to zero if infinitely small grains were available.

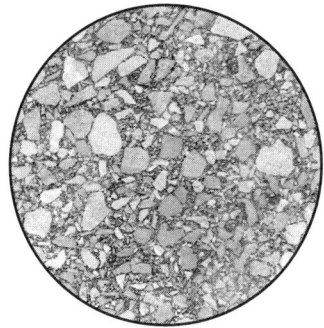

Figure 10.4
Cut of a block of concrete revealing angular grains in a large range of sizes.

Modern high-performance concretes owe their somewhat accidental success to the integration of silica powders a few microns large. This
refuse from industrial metallurgy had to go somewhere, and it went from
being waste to being a commerical product! The fine powder may fill
the smallest pores between grains of matter. The material's porosity, that
is, residual water content after drying, is reduced to a low percentage.
Since the voids are weak points of a porous material, adding silica powder
increases the mechanical strength. Here are a few relevant numbers: ordinary cement can make concretes able to support a load of 20 MPa (one
MPa corresponds to 100 tons per square meter). At the other end of the
spectrum, *high-performance concretes* may display a compressive strength
up to twenty times stronger. At the same time, their tensile strength is
on the order of the compressive strength of ordinary cements, which
have a weak tensile strength (as do porous media). The new Museum
of European and Mediterranean Civilizations in Marseilles (designed by
well-known architect Rudi Ricciotti)—with its elegant, "netted" walls—
provides spectacular examples of this high-performance material at work.
The slightly curved walkway 120 meters long connecting the museum
to the fortification at the entry of the well-known Old Port has a deck
only a few centimeters thick, thanks to new concrete materials incorporating fibers. It is a cheaper construction than a metal walkway; inasmuch as less expensive and more readily available materials are used,
it has a positive ecological impact. As noted, the issue is high performance, not simply high strength. Such concrete can be transported in
liquid form up to the top of high buildings. Thanks to polymeric adju-
vants, concrete of this kind remains highly fluid during the construction process. Finally, because of reduced porosity and a low quantity
of residual water in the pores, concrete is less sensitive to the effects of
chemical degradation, which reduce durability (as would be the case
for the new museum in Marseilles if it were directly exposed to saline
sea sprays).

CEMENT

The mechanical performance of concrete depends on that of the cement.
In its anhydrous form, cement is a fine powder that results from curing

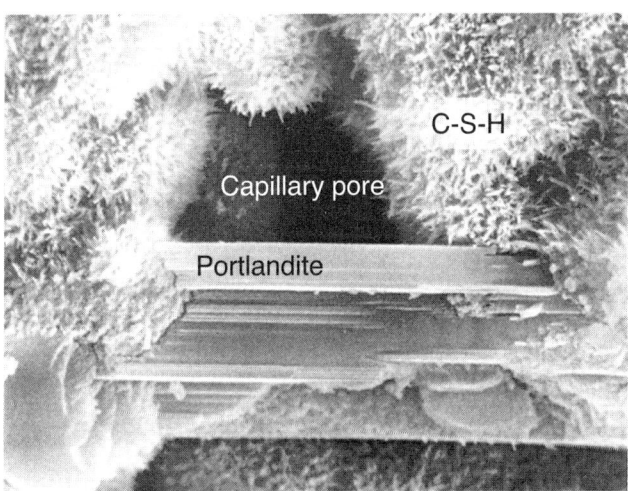

Figure 10.5
Internal structure of a cement composed of small particles of calcium silicate hydrate (CSH) that are elongated and entangled. They create bonds between concrete granulates. Portlandite plays mainly a space-filling role.

a mixture of clay and limestone. Calcium silicate hydrates are the main hydrates produced when water is added to powdered cement. Hydration is a process of dissolution and precipitation that makes the cement paste harden over a period that can last from a few hours to a few days. After it sets, the medium comprises long particles a few nanometers in diameter, pores, and water (figure 10.5). Each particle is structured in sheets composed of a double octahedral calcic layer held between two layers of silica tetrahedra. Water forms part of this structure and contributes to rigidity by binding the sheets. The cohesion of cement simultaneously arises from the internal cohesion between sheets and the cohesion of particles in the hydrated cement. As such, hardened cement may be said to be a nanogranular material.

NATURAL CONSOLIDATED MEDIA

SEDIMENTARY ROCKS

Undoubtedly, most examples of cohesive granular materials are found in sedimentary rocks. Under the effects of time and extreme conditions of temperature and pressure in the subsoil, layers of accumulated sediments cover about 50% of French terrain. The layers of successive deposits visible on rock faces bear witness to the history of these deposits and the deformations they have undergone over the ages. A prime example is sandstone. These quartz grains are easily recognized under a microscope; by scratching, they can be extracted one by one. Porous sandstone is often a reservoir of petroleum; as we will see in chapter 11, knowledge of the geometry of pores is necessary to optimize its recovery.

BACTERIAL AGENTS

Concretes do not represent a prerogative of human beings. *Sporosarcina pasteurii* is a small bacterium that researchers have been keenly studying for the last dozen years. As it breaks down urea, *Sporosarcina pasteurii* creates a precipitate of calcium carbonate that cements grains of sand together, transforming soils into veritable rocks that can be as hard as sandstone. The process holds great interest because of its incomparably low impact on energy. What's more, these bacteria occur naturally in the soil and pose no problem for the environment, whereas products traditionally used for reinforcing soils (e.g., epoxy resins) are often highly toxic. The potential applications are numerous, including to repair fractures in ancient monuments and stabilize sloping terrain or soils subject to liquefaction (which become especially unstable in the event of an earthquake). Certain architects have even considered using these bacteria to build directly with desert sand! As crazy as it sounds, the principle would resemble the idea behind 3D printers: boring small holes at precise locations and injecting the bacteria. Once the soil transformed into rock, the surrounding sand would be removed to reveal the structure.

The world of animals and plants also uses materials with a structure very close to our concretes. Termites (*Macrotermes bellicosus*) that build earthen cathedrals are masters of working with an organic cement made from droppings and undegraded sugars. To construct mounds several meters tall, with a complex architecture that can withstand wind, sun, and rain, requires material that's high-performance and easy to use.

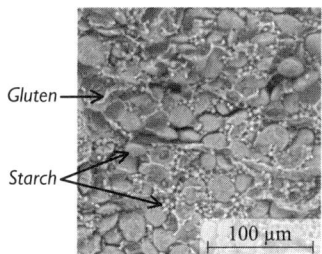

Gluten

Starch

100 μm

Figure 10.6

"Organic concrete": wheat endosperm, viewed under a scanning electron microscope, composed of small and large starch granules (grains) and a net of proteins called "gluten."

Researchers are highly interested in the properties of the termites' binding agent and the ways it combines with particles of soil. Although it is less resistant than modern concretes, the material has the advantage of being biodegradable and it can be produced at ambient temperature, which means low energy consumption.

WHEAT

Grain reserves primarily contain more or less spherical starch grains stuck together by an assemblage of proteins called *gluten* (from the word *glue*) (figure 10.6). In fact, this is a highly durable biological concrete. Just try taking a popcorn kernel and breaking it between your fingers; even crushing it with your teeth is difficult! A bit further on, we will examine the mechanics of fragmenting and extracting flours, semolinas, and bulghurs used the world over for feeding human beings and animals.

THE MECHANICS OF GRANULAR COMPOSITES

For all the granular materials discussed in this chapter, the quality of adhesion is controlled both by the quantity of the matrix (consisting of, for example, cement, gluten, industrial residues) and the way the matrix adheres to grains. These parameters influence the *elastic moduli* of the material and its *fracture* threshold. The following presents a few aspects of the mechanics of such heterogeneous media.

TYPES OF FRACTURE

Researchers have conducted experiments at the University of Montpellier in order to understand the effects of different kinds of fracture. They used spheres of expanded clay as a granulate, and tile adhesive for the cementing matrix. Silicone coatings were applied in varying quantities to grains to control the adhesion between grains and matrix. Researchers observed three different types of fracture as a function of the amount of the matrix and its adhesion with the grains. The experiments show that under the action of an increasing compressive stress, the samples fail at a stress threshold that depends on both parameters. For weak adhesion, cracks appear and spread at the grain-matrix interface (figure 10.7, type 1), but they circumvent grains without damaging them—which is exactly what happens when you break a chocolate bar with nuts! Such material is *friable*, or brittle, whatever the amount of matrix, provided that the grains fill the volume well. In contrast, if the volume fraction of grains (packing fraction) is low with respect to matrix volume, the mechanical strength depends more on the mechanical properties of the matrix than on those of interfaces.

When adhesion between grains and matrix is good, two regimes of fracture are observed, depending on the amount of the cementing matrix. If the matrix takes the form of low-volume bridges between grains, localized abrasion is observed (type 2) on the surface of particles. For amounts extending from the matrix's percolation threshold up to its saturation of the space between particles, cracks will propagate across the grains, provided that the grains are not much stronger than the matrix (type 3).

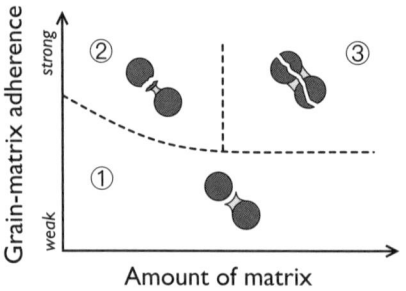

Figure 10.7

Map of the three types of fracture as a function of the amount of cementing matrix and the grain-matrix adhesion.

MILLING

These ideas can be applied to milling wheat. Several thousand kinds of wheat exist, and each has its specific features and genetics of its own. These differences are compounded by the high level of variability that environmental conditions impose on plants. If soil is rich or depleted, if there's too much or too little rainfall, the composition of grains will vary, thereby changing wheat's mechanical strength appreciably. In order to understand such variability, one must consider the resistance of a given wheat grain and conduct analysis at the microstructural level, between gluten and starch granules.

Atomic force microscopy (AFM) has been used to evaluate the strength of gluten and starch and the interface between them. If stresses are strong in a zone of weak resistance—in the vicinity of a pore or fault, for instance—it will represent a "weak link," where fracture can be triggered. Often, such weakness occurs at interfaces between components. As in the model experiments just described, adhesion at the level of granules and gluten plays a major role in the overall strength of wheat grain. Research has shown that certain proteins, *puroindolines*, weaken this interface by reducing surface adhesion. To map out the factors that influence grinding quality, the number of damaged granules is represented along two axes: one represents the percentage of gluten filling the space between granules, and the other the amount of the puroindoline at the interfaces. The three regimes of fracture described earlier (figure 10.7) are shown for wheat endosperm.

At the same time, weakness can also be found at the level of components. Gluten is a net of intertwined polymers that can undergo a large deformation without presenting too much resistance before breaking. In contrast, starch granules have a "sandwich"-like structure where crystallized and amorphous layers alternate. This structure can withstand significant stress, but no large deformations. During the grinding process, starch granules (especially large ones) are broken down. The amount of damaged starch is an important factor for the quality of breads, biscuits, and noodles, for instance. One of the main reasons is that a damaged starch absorbs water several times its weight whereas undamaged starch will absorb less than half. In breadmaking, a small amount of damaged starch is necessary to increase the dough's absorption capacity; but if there's too much damage, the dough will become sticky and slack and prove difficult to work with.

STRESS TRANSMISSION

In granular composites that contain a high volume fraction of grains, forces are transmitted through the matrix as well as through grain contacts. This transmission from grain to grain is similar to the *force chains* mentioned several times throughout this book. Figure 10.8 shows two examples of stress chains. In the first case, matrix volume is low, and forces are carried by contacts between grains more than anything: long force chains are evident, as in uncemented granular media. In the second case, the matrix fills all the space between grains and also bears an important part of the stresses applied to the overall system. For this reason, forces between grains aren't as strong, and long stress chains no longer appear.

The role of force chains for concentrating stresses in composites is less well known than the effects of pores. In a porous or cracked medium, the stresses concentrate around pores and cracks. A material that is subjected to external shear stress or tensile stress starts to yield by fracturing in

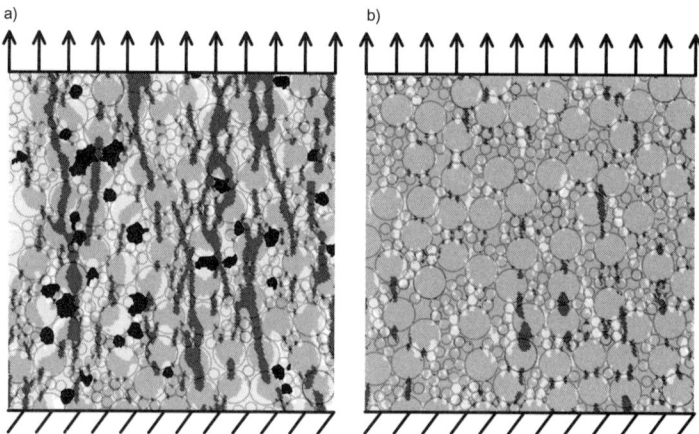

Figure 10.8
Simulated tensile stresses (shades of gray represent the intensity of tensile stresses) in a granular composite with pores (in black). The pores present in simulation a) have been filled in b) where stresses are distributed more homogeneously throughout the medium.

zones where stresses are highly concentrated. In a composite, the concentration of forces between grains in the form of force chains can play the same role and lead to failure. In this respect, the matrix plays two distinct roles: 1) as a glue for sticking particles together, and 2) by filling space between grains, which permits partial redistribution of the load between contact network and matrix. The latter contributes to reducing the inhomogeneity of stresses and increasing the material's mechanical strength.

In this chapter, we looked at examples of composite materials that incorporate grains and granular media consolidated by adding binding material between particles. The physical mechanisms governing the strength of these materials are complex and continue to be the focus of experimental study and new modeling approaches. The volume fraction of the binding matrix and its adhesion with the grains are important, but other parameters also prove significant. For instance, the distribution of particles in space and the degree of disorder can modify mechanical strength. Depending on their distribution, the gaps between grains can be larger or smaller and allow stress chains to form and percolate—or not—across the medium. It is well known that the presence of rigid grains in a continuous medium increases mechanical strength: grains can stop the propagation of cracks or divert them by rendering the cracks more costly in terms of energy required. By controlling the spatial distribution of consolidated grains, then, it is possible to manufacture composite materials with new properties.

11

FLUIDS IN GRANULAR MATERIALS

We might say that the earth has a spirit of growth; that its flesh is the soil, its bones the arrangement and connection of the rocks of which the mountains are composed, its cartilage the tufa, and its blood the springs of water.

Leonardo da Vinci

Up to this point, the fluid present between the particles of granular media has remained relatively calm—stuck between the grains of a sandcastle, then moving a little when it is a matter of modifying its packing fraction. It's time for it to flow! From the perspective of grains themselves, we will now examine how fluid flows in a porous medium, which it may modify by its action. The next, and final, chapter will take the opposite approach, exploring how a collection of grains is swept along by a moving fluid. The two points of view complement each other. Needless to say, these two perspectives overlap in the case of a medium in which both the fluid and the grains are in motion.

The flow of a fluid through the pores of a fixed packing of grains represents a vital subject of investigation, and the name of Henry Darcy will remain associated with the endeavor for the ages. A civil engineer, Darcy was in charge of providing the city of Dijon with water. His technological innovation made it the first city in France to have running water in every building as early as 1850. To this end, he studied the transport of water in soils from nearby waterways and, in the process, identified the fundamental law of permeability of porous media, now known as Darcy's law.

Porous materials are not restricted to the channels between the grains of a granular medium. A wide variety of porous structures exists. Any speleologist who has explored grottoes and sumps will have much to say about all the work performed by water circulating through porous or cracked rock structures. For ease of reference—and to avoid a veritable litany of all the existing porous media—we'll use the familiar bag of marbles as a model.

GRANULAR MATTER IS FULL OF VOIDS

A little-known fact is that a sand dune can serve as a reservoir; it is capable of holding water up to 30% of its volume. Porosity ϕ is the fraction of voids in a granular packing; it represents the complement to the packing fraction C, so that $\phi = 1 - C$. However, this fact tells us nothing about the winding and labyrinthine geometry that a fluid, or a particle, will follow as it moves in the pore space—nor does it shed any light on the properties of fluid flow in the pore space. The latter also depend on the properties of fluids, whether static or flowing. In this chapter we will distinguish between *saturated media*,* where only one fluid phase is present (e.g., a *water table*, or soil entirely filled with water), and *partially saturated* media (as in the case of soil closer to the surface, which contains a mixture of water and air).

The image on the right side of figure 11.1 represents the space of a void between four soccer balls of the same diameter that form a tetrahedron. We can picture a flow network in schematic terms as a central, spherical

a)   b)

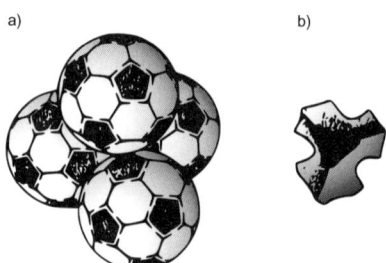

Figure 11.1
a) The tetrahedron is the elementary structure for a packing of spheres. b) Pore space on the same scale.

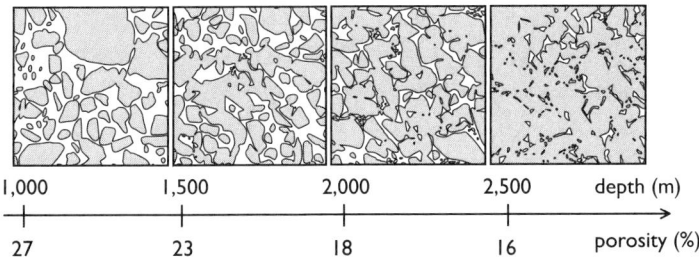

Figure 11.2
Visualization of a cross section of the pore space in Massillon sandstone taken at different depths. The pressure effect from the upper layers (lithostatic pressure) leads the pore size between grains to decrease with increasing depth without significantly changing the size of grains themselves, which get strongly deformed. Deep down, a large number of pores are no longer connected.

zone connected to neighboring pores by tubes of a smaller diameter—or, alternatively, as an ensemble of cylindrical tubes joined together by pockets. The greatest part of porous volume is contained in the pockets, but the fluid's transport properties will largely be controlled by what happens in the narrower tubes or *throats*, representing the small cross sections between pockets. Bear in mind that this is a crude analogy.

When studying sintered materials, one can characterize pore geometry by taking cross sections. A model system often used for this purpose consists of glass spheres that have been sintered at different temperatures and for different durations. At low sintering rates, porosity approaches that of an unconsolidated packing (30%). As sintering proceeds, it decreases considerably, until many pores are eliminated (see the difference between figures 10.3c and d). A continuous path no longer exists for fluids with a porosity below 10%. A second example is sandstone, with samples taken at increasing depths (figure 11.2). In this case, sintering is due to the pressure effect of the upper layers. As pore size decreases with depth, the path between pores becomes increasingly tortuous. We recognize that here the notion of percolation, presented in chapter 6, applies.

PERMEABILITY

The time has come to look at flow in a porous medium. The words *permeable* and *porous* are often confused, but their meanings are quite different. *Porosity* is the volume fraction of empty space. *Permeability* is a fluid's capacity to flow through porous material in response to pressure difference (just as *conductivity* refers to the capacity of charges to move in a medium under the effect of a difference in electric potential). For instance, Swiss cheese is very porous, but the holes are not interconnected: thus, the medium is impermeable. The combination of high porosity and impermeability is desirable in construction materials, to ensure lightness and good thermal, or phonic, insulation. For another example, take two vessels filled in the same way, but with spherical balls of different diameters. The porosity is the same in either case, but permeability is much higher for the large pores between large spheres than it is for small pores between small ones. If you make espresso with coarsely ground beans, the flow rate will be faster … but the coffee will be weaker!

FLOW IN A TUBE

The Poiseuille equation connects the flow rate of a viscous fluid in a straight cylindrical tube to the fluid's pressure difference between its two ends (figure 11.3a). To calculate the permeability of a granular material in approximate fashion, one can picture an ensemble of small cylindrical and winding tubes joining the ends of the porous material in which the fluid runs (figure 11.3b).

The Poiseuille Equation

Jean-Léonard Poiseuille (1797–1869), a doctor and a physicist, examined the effects of viscosity in blood flows in narrow tubes and formulated a law that expresses the relation between flow rate Q in a cylindrical tube of diameter d and pressure drop per unit of length $\Delta P / L$ between its ends (figure 11.3a) in a continuous flow regime. Viscosity η is a characteristic of the fluid. The flow rate is proportional to $(d^4 / \eta) / (\Delta P / L)$. This rapid variation as the fourth power of the diameter, and not simple proportionality to the cross section d^2 (as would be the case for calculating the electrical current in a conductive wire), may seem surprising. However, it stems from variation of the velocity gradient between the maximum speed at the center of the tube and its null value at the edges. This causes a great pressure drop for the tiny tubes. People prone to frostbite at the tips of their fingers know that blood flow is quite slow at the level of capillary vessels!

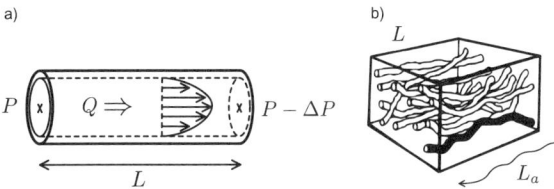

Figure 11.3

a) Poiseuille flow of a fluid in a cylindrical tube of radius R in laminar regime. Velocity is maximal at the center and zero at the edges. Thus, the total flow rate Q diminishes very quickly with its cross section. b) Flow in a model porous medium made of independent tubes whose tortuosity equals L_a / L. Total flow is proportional to the difference (or fall) of pressure ΔP as in the case of Poiseuille flow.

DARCY'S LAW

The Poiseuille equation expresses the proportionality between the mean speed of flow in a tube and the pressure difference between its ends. Such linearity persists for a porous material if one takes the average of the flow rate in a volume much larger than that of a pore. *Darcy's law* states that, for a porous material of cross section S and length L, with pressure difference ΔP between its ends (figure 11.3a), the volumetric flow rate Q / S per surface unit S for a liquid of viscosity η is equal to (κ / η) $(\Delta P / L)$. This formula resembles that of flow in a tube. κ, which stands for *permeability*, is a characteristic of the geometry of the porous medium. Whereas porosity is expressed by a number, permeability has the dimension of squared length: its unit is the *darcy*. For a collection of tubes, the permeability would simply scale as the square of a single tube's diameter d^2. This law is valid only for flows controlled by viscosity at a low *Reynolds number.**

Figure 11.4 provides an example showing how permeability varies with porosity in a series of samples of sandstone. If, as before, we look at samples whose porosity decreases with sintering or densification, we observe that the decrease of permeability is more pronounced at low levels of porosity. Here, more and more regions appear that are porous *but also* impermeable: paths without exit for the fluid (as in the deep-lying sandstone in figure 11.2). In order to describe this domain, it is necessary to take into account more detailed features of geometry—as we did in chapter 6, in the context of mixtures of conductive and insulating grains. (The pioneering work of Hammersley on *percolation* [presented in

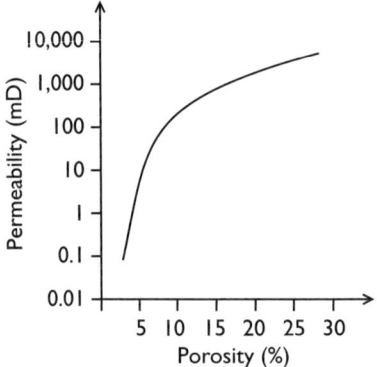

Figure 11.4
Variation of air permeability of Fontainebleau sandstone as a function of porosity. Note the lower levels with decreasing porosity—with a very rapid drop below 10% due to closed porosity.

chapter 6] was initiated by considering the clogging and loss of permaeability of gas masks as porous media.)

Permeability can also be calculated with three geometrical factors, following the model presented in figure 11.3b: the *porosity*, the volume fraction of tubes; the *diameter* d of channels; and the *tortuosity*, which represents the fact that traversing pores of a granular medium means following twisting paths. The tortuosity can be estimated rigorously, on the basis of direct tomographic analysis of the pore network, or by electrical measurement that involves filling the porous medium with a conductive liquid. The resistance of the soaked porous medium will be larger, the more the paths taken by the current are winding.

CHARACTERISTIC LENGTH OF A POROUS MATERIAL

The permeability of a porous medium is measured by a squared length. The radius R of the spheres of our marble-pack model might come to mind. However, the soccer-ball model (figure 11.1b) indicates that the size of throats is closer to $R / 7$. When squared, this already represents a reduction factor of $1 / 49$! Since the size of naturally occuring grains ranges from one to a few hundred microns, employing the legal unit of permeability—a square meter—would yield extremely small numbers. The *darcy*—about one micrometer squared—therefore provides a convenient subunit.

A second length scale emerges if the material is sintered, as in figure 10.3. The size of grains does not vary greatly, but pore radius decreases with sintering; it can be much smaller than the radius of grains. Once again, an indirect method allows us to define the relevant length, employing the *critical radius* r_c presented here.

Critical Radius

Here's a picture for understanding the idea of critical radius, as it applies to all kinds of disordered networks and systems interconnected in highly inhomogeneous fashion. Consider a lattice of roads, with freeways, highways, byways, and dead ends. Suppose that an engineer has set up the elements of this network at random. Let's focus on the connection between two large cities (Paris and Marseilles, for instance, at opposite ends of a French map). The permeability of highways with three lanes—elements with good conductivity—does not restrict the average flow rate. Traffic on parallel routes, smaller roads traveled by tourists that drivers in a hurry avoid, doesn't restrict it, either. What matters are the most unfavorable elements (two-lane traffic, for example, or bottlenecks due to accidents) on the most important roads for transport. In the same way, the critical elements in a pore space are the narrowest pores in the "right" paths the fluid needs to take. This leads to the notion of critical pore radius r_c. In the section "Pore Invasion," we will discuss its measurement by means of mercury porosimetry.

EVALUATING PERMEABILITY

On this basis, we have the necessary ingredients for determining the permeability of a bed of grains (which is essential for water hydrologists and petroleum engineers): tortuosity as determined by electrical measurements taken by filling the porous material with a liquid of conductivity G_f and by a determination of the length of a critical radius r_c. When a broad range of pore sizes exists, r_c controls the flow rate of the fluid and electrical current simultaneously, because these two quantities use the same percolation network, even if the law governing the transport locally isn't the same for both phenomena.

Because of this geometrical similarity, permeability and electrical conductivity will simply be proportional. This leads to establishing a law: $\kappa = \alpha\, r_c^2 / f$; α represents a numerical factor. Permeability is proportional to the square of the critical radius and to the conductivity G_p of the pore-filling medium, which varies as the inverse of a parameter called *formation factor* G_f / G_p providing an estimate of the variation of permeability due to reduced and tortuous geometry. This empirical result was proposed by two physicists working for Exxon, A. J. Katz and A. H. Thompson, who verified it for a large variety of artificial and natural porous media by measuring the conductivity G_p of the porous medium filled with a fluid of conductivity G_f, the critical radius, and the permeability. Besides providing a means for correlating the properties of a rock, this affords striking experimental proof of the concept of *critical path*, generalizing the idea of percolation and applying it to multiply connected systems with paths displaying a broad range of local transport properties.

UNSATURATED MEDIA

Porous materials are often filled by several fluid phases at once: the mixture of air and water in a soil (except after heavy rain, which saturates it) is one example. Another example is an oil-reservoir rock containing a mixture of air and petroleum or tars. Such pairs of fluids are characterized by interfaces, menisci between air and water (or oil and water), and the ways that fluids wet the pores. Chapter 9 familiarized us with forces of capillarity, which are responsible for the cohesion of a sandcastle. Capillarity plays a major role for porous materials, both because of the interface between the two phases of a given material, and because of these fluids' contact with pore surfaces.

IMBIBITION AND DRAINAGE

When a glass tube open on both ends is stuck vertically into a container of water, the water inside the tube will rise to a height h that is larger, the smaller the tube's radius r is. The meniscus at the upper surface of the water is practically a half-sphere open upward (figure 11.5a). This is a case of *imbibition* of water—like when we put a cube of sugar on the surface of coffee. One says that "water wets glass" in order to express the fact that water's contact angle with glass is near zero: water has a positive affinity for glass and spreads readily on it.

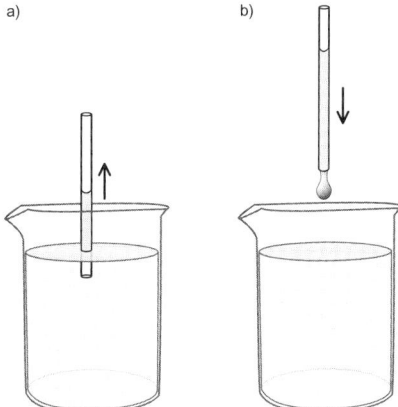

Figure 11.5
a) For a fluid that wets glass, capillary rise is inversely proportional to the tube's diameter. b) Schematic depiction of a tube draining under the effect of the water column's weight. A (delicate) mechanism balances the weight of the column with the effects of the capillary forces in the two menisci at top and bottom.

If we remove the tube completely once it's filled, the water will flow downward (figure 10.5b) under the effect of its weight. This is an instance of *drainage*. The simple phenomenon can be described in terms that, while more complicated, will ultimately prove to be useful: the air above the water plays the role of a nonwetting phase. It does not interact much with the walls and has an opposite effect with respect to the water interface (which wets the glass). The pressure of the liquid at the top of the column is lower than that at the bottom by an amount corresponding to the water height in the capillary network. Just below the upper meniscus, then, water is depressed relative to the air. We can express the matter as follows: "Air under pressure, which does not wet glass, expels the layer of water, which wets glass."

PORE INVASION

A small drop of mercury placed on a plate of glass forms a spherical drop, expressing the fact that mercury does not wet glass. If a vertical tube is plunged into a layer of mercury, the level of mercury in the tube will lie below the surface level outside. It is the opposite of what happens with

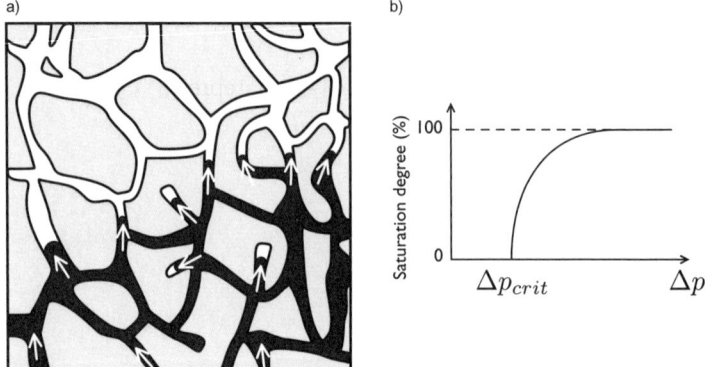

Figure 11.6

Injecting a nonwetting fluid like Hg (in black) into a porous material (schematized by interconnected tubes) demands a difference in pressure between fluid injected (Hg for example) and fluid expelled (in white). This difference, given by a value Δp, is inversely proportional to the radii of fluid interfaces. At the pressure percolation threshold, a continuous (black) path will connect the top to the bottom of the medium as shown in a). The graph b) shows how the penetration of mercury (saturation of the pores) varies in a porous medium as a function of applied overpressure.

water. The *mercury porosimetry* experiment involves forcing mercury to enter one side of a porous material under pressure. Because, according to the Laplace-Young law, mercury hardly overcomes the adverse capillary forces in very small pores, only the largest ones will be invaded (figure 11.6). This number of pores increases with injection pressure, and the mercury makes its way in a series of long, tortuous paths, with a front advancing in the medium. In the process, it leaves behind surrounded regions of volume with pore diameters that are too small to be invaded. There are also weak points, where passage needs to take place. The picture calls to mind the invasion of Normandy in June 1944 with its "Falaise Pocket," where the German forces remained surrounded by Allied forces on their way to Paris. In this analogy, the weakly active but necessarily accessible pores are like the bridges on the river Seine; their bombardment would have stopped the progression of the Allied forces. Thus, the idea of "invasion" naturally introduces itself in the concept of *invasion percolation*, which characterizes the penetration of mercury into a dry porous medium. The number of channels invaded increases until a critical pressure is reached and a continuous path appears across the pore network. The pores of radius r_c are the last to be filled when the threshold of invasion percolation is reached—they represent the smallest of the large, active channels. This measurement of r_c makes it possible to establish the correspondence between permeability and conductivity in porous media. Chapter 6 already illustrated the general problem of percolation for mixtures of conductive and insulating spheres, with an increasing number of conductive spheres; thereby, the percolation threshold corresponded to the proportion of conductive spheres large enough to make the mixture conductive. In invasion percolation with mercury, the percolating phase is characterized by a large number of occupied channels—or, equivalently, by a large saturation degree.

What's more, the law of invasion as a function of pressure provides information about the distribution of pore sizes. Laplace-Young's law, on the basis of pressure measurement and the mercury's surface tension, allows us to determine the characteristic length scale of pores: the pressure at which the system first interconnects gives the critical pore radius r_c.

SOIL IS A RESERVOIR

This statement is true even for deserts, we should remember, provided that water is sought deep down, as by plants whose roots can plunge dozens of meters. A dune is a water reservoir that we are still unable to tap!

Models of capillarity enable us to account for physical effects of imbibition when a soil is wet, or for drainage as it dries. Soils are heterogeneous and constituted by an ensemble of *horizons* that change with depth. Capillary forces hold water in these bands, forming films and menisci that depend on the physical chemistry of media and their surfaces. However, we should note that the roots of plants cannot mobilize all the available water. Thus, the amount of water may be significant but trapped by sheets of clay (see figure 9.14); roots can collect only a small fraction of the water charged with mineral salts vital for the plant. Silt, and then sand, has water that's more readily available—due to the larger material's large pores. Pedologists, or soil specialists, speak of *capillary potential* to characterize the way water can be mobilized by plants; high potential corresponds to narrow pores (see figure 9.8). Water cannot be collected beyond a critical value of capillary potential. Working the soil lowers this potential by creating larger voids.

MISCIBLE LIQUIDS

MOLECULAR DIFFUSION ...

We have just described the behavior of fluids when several immiscible phases inhabit a porous medium simultaneously. What happens when the fluids are mutually soluble? What about when there's just one liquid phase, but its composition changes continuously throughout the space it occupies? If we put a drop of milk in a glass of water, it will spread. If the density of the two liquids were the same, the motion of spontaneous agitation displayed by droplets of fat suspended in the water would govern the slow process of diffusion. The radius of the milky stain would grow as the square root of time; roughly speaking, the drop will spread only ten times more if we wait one hundred times longer. This is indeed a slow process and there is a more efficient way:

... AND DISPERSION

In porous materials, the fluid flow takes the place of classic diffusion and it disperses pollutants efficiently on a large scale (figure 11.7a). This mechanism is extremely important; it comes into play, for instance, when a water table is subjected to local pollution, raising the questions of how far will the pollutant spread, starting when, and for how long?

Picture a model medium in which a fluid is flowing. As in experiments of molecular diffusion, let's add a pollutant locally for a brief spell. It will spread at the average speed U of flow (figure 11.8). If we followed the pollutant's trajectory as it is transported at this average speed U—as if it were sitting in a little boat moving at its average velocity on an agitated waterway—we'd see the flow move locally faster or slower than the boat, with a relative speed of zero. These local variations of velocity are due to the random paths the polluted fluid follows between grains of average size d at speed U, producing a dispersion of the order Ud. Note that

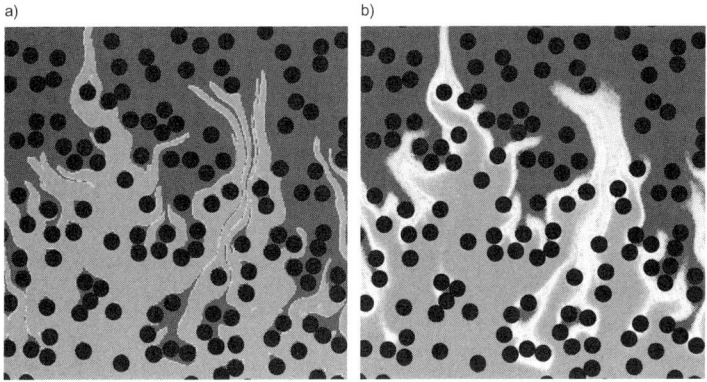

Figure 11.7
a) Flow of a gray fluid from bottom to top expelling a darker fluid in a two-dimensional porous model (circles) without diffusion. b) The same flow, with diffusion creating transverse broadening (in white).

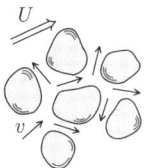

Figure 11.8
The dispersion process may be illustrated by a random walk model. U is average velocity and v is local velocity.

thermal agitation is not involved here: the effect is due to the disorder of the flow.

If molecular diffusion is very low, zones of fluid where the flow velocity is weak, near the edges of porous material or in crevices, will take time to return to zones of rapid flow; dispersion effects will make themselves felt over a much longer period when molecular diffusion is small (this is the opposite case of simple diffusion). Roughly speaking, small molecular diffusion implies large dispersion. This mechanism, known as *Taylor dispersion*, occurs especially in *Poiseuille flow** along a tube in the presence of localized heterogeneities. Consider turning on the hot water in a shower: Taylor dispersion occurs during the time it takes for water to reach its final temperature in the pipe initially filled with cold water.

MODIFYING PERMEABILITY

Until now, we have taken into account the way that a porous medium can be altered by pore flow. The fluid circulating there may contain sediments in solution or grains that deposit and block pores. This is the mechanism of the very slow consolidation of sedimentary rocks. In order to reduce leaks, blocking pores will be essential for storage of CO_2 in depth. In contrast, technologies for hydraulic fracturing and acidifying soils aim to enlarge pores that are too small, in order to extract the petroleum or gas they contain. Finally, we will also examine how a porous filter's permeability is modified by the grains it accumulates.

PLUGGING POROUS MATERIAL

An especially important situation concerning the evolution of a sediment's permeability involves the slow deposit of substances in the form of ions in solution, which cover the grains and consolidate the material. This geological process, *diagenesis*, occurs over extended periods of time and leads to the formation of sedimentary rock. Figure 11.9 shows the same sandstone before (figure 11.9a) and after (figure 11.9b) a deposit has occurred on grains; it calls scaling pipework to mind.

a)   b)

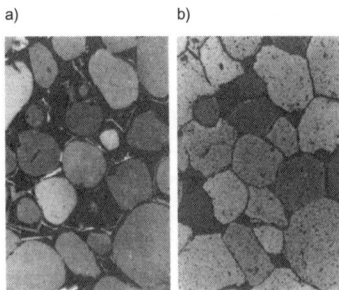

Figure 11.9
This figure shows the same section of highly cemented sandstone. a) Cathodoluminescence enables the sandstone to be seen as it was before the deposit of a cementing material. b) Its present state, with the permeable zone in black, which is greatly reduced.

UNCLOGGING POROUS MATERIAL

It's important to be able to restore the permeability of hydraulic systems that have been plugged. This is the case of depth filters that have lost their porosity. As we have seen, the theory of percolation was developed while studying how gas masks became clogged. Restoring the permeability is simply a matter of changing the filter. But in process engineering, the cost and the time lost in such an operation may can be avoided, since one may simply reverse the flow for a short interval of time to clean the filter.

Another means of cleaning pipes involves injecting a liquid containing a product that attacks limescale buildup but not the conduits themselves. When the porous material under consideration is a network of interconnected pipes, the task poses a challenge. As fluid passes continuously from one end through the filter, it uses various pathways at the same time. In consequence, some pipes experience a greater amount of fluid passing through. Over the course of time, these channels will be more favored, flushed more quickly, and so on. Ultimately, one would end up with a single large open channel where all the fluid passes, while the rest of the pores remained clogged! Hence, new mechanisms or materials must be sought to obtain a different result—for instance, fluids whose viscosity increases with flow rate. An experiment you may have performed as a child is building a dam on the beach to hold the water filling it from the top as the sea recedes. Finally, a small crack on or in the dam will have let the water in, leading the edifice to collapse. The phenomenon is *piping*: the water, filtering in a very heterogeneous way through a barrier on or underneath the earth, progressively increases the local flow rate and the transport of grains outward—the way an animal makes its burrow—until it is destroyed.

FILTRATION

The transport of solid particles in a porous medium and the medium's capacity to block particles via *filtration** represent a major domain of application for both porous materials and filters.

In general, filtering operations that stop solid particles in a fluid entail clogging, whose effects need to be minimized. Objects of practically any size may be blocked by means ranging from grids and sieves all the way

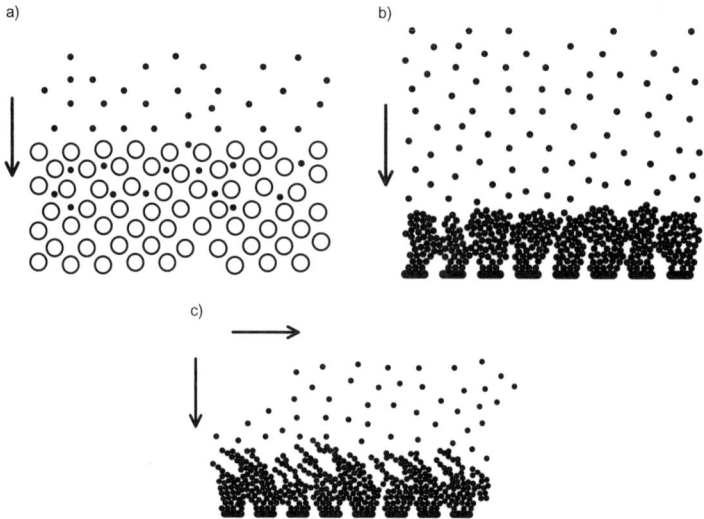

Figure 11.10

Three modes of filtration: a) filtration in a porous volume involves the transport of small particles (black dots) between pores; b) in transverse filtration the particle-laden fluid comes from above and progressively builds up a filtration cake; c) in tangential microfiltration, the fluid comes from the left and is filtered in the bottom part of the filter.

up to ultrafiltration membranes. For particles of small size, and for chemical and biological purposes in particular, the surface of the pores in a filter plays an active role. Here, we will limit discussion to basic principles and stick as closely as possible to granular media.

Depth filtration (figure 11.10a) employs a very thick granular medium, which may be natural (in the case of soil) or packed in a filter column. The filter's pores are larger than the particles it is intended to stop, to avoid a deposit forming on the surface where the liquid passes through. This type of filtration is employed to purify highly dilute suspensions (less than 0.5 gm per liter) of small particles (with a diameter below 10 μm); the porous material consists of grains with diameters between 500 and 1,000 μm. This method is in broad use for treating polluted water. The way the particles flow is governed by the porous medium's structure. However, this structure evolves in a very inhomogeneous manner as particles deposit in the filter, since layers close to the point of entry are saturated more quickly. Throughout the operation, the permeability of the filtering medium diminishes; this evolution of flow conditions make it difficult to analyze depth filtration fully.

Let's examine how a single grain carried by a clean fluid encounters the porous medium, and the processes that can bring it into contact with pore surfaces. Our particle is subject to hydrodynamic forces that tend to displace it along a flow line of pure fluid. At the same time, however, a number of factors may prompt it to change its trajectory. At the level of narrow pores, current lines tighten and draw close to the surface of grains. In consequence, the particle may hit a grain of the porous medium if the flow line passes at a distance from the side of a grain that is smaller than its diameter. We also need to take into account inertial forces, which tend to make the particle move in a straight line. (This effect is especially sensitive to abrupt changes in the curvature of the flow lines: it's the same as when a driver takes a turn too fast and hits the guardrail). Finally, Brownian diffusion influences our solid particle if it's small enough: a transversal motion of random diffusive agitation is added to the motion globally oriented along the flow direction. Under these circumstances, effects of physico-chemical entrapment play a key role for small grains.

While large grains prove sensitive to inertial effects and small ones to thermal agitation, particles of intermediate size pass through more easily. The corresponding size of filters, of the order of a dozen microns, is precisely—and unfortunately—that of particles of cigarette smoke. Whether a particle that enters into contact with the porous medium is captured or not by the porous medium depends on other physico-chemical forces. The interfaces between particles and liquid carry electrical charges, which result in repulsive forces whose radius of action, on the order of 1 μm, depends on electrical charges in the solution; these forces manifest themselves at a longer range when there are few ions present. (Adding ions will reduce this range or even create attraction between the particles and larger objects, which is a means of accelerating sedimentation in order to clarify polluted water.) Likewise, van der Waals forces of attraction act at very short range, on the order of thousands of angstroms; the attraction they exert can lead particles to adhere to the surface of the grains of the filter.

So what happens when an ensemble of particles flows through a porous medium? Their conditions of transport and retention depend, in a complex manner, on all the forces previously listed. In fact, two extreme situations can be distinguished. The first concerns *hydrodynamic chromatography*, whereby the presence of porous material modifies the particles' speed of motion with respect to that of the liquid, without filtration occurring. If particles of varying sizes are injected into the porous medium, we can see that the larger they are, the faster they will traverse the medium. This unexpected effect provides a method for separating microscopic particles according to size. Remember that the profile of a liquid's flow velocity is parabolic in a capillary tube (figure 11.3b). By the same token, in a porous medium flow velocity is zero at the sides and reaches a maximum at the center of pores. Jamming effects and electrical forces repelling them exclude particles from the vicinity of the pores' surface: accordingly, they don't "see" the liquid's smallest flow lines near the edges and move at a speed faster than the liquid's mean flow velocity. A small particle will move less quickly than a larger particle because, given its Brownian motion and size, it is able to draw closer to the sides where the flow is slower. In the realm of hydrodynamic chromatography, particle flow is governed by jamming and repulsive electrical forces, whose combined effects prevail over those of van der Waals forces of attraction. The second extreme case concerns *entrapment*. Above a certain threshold size, particles are all trapped inside porous material—this is the domain of filtration, which proves more efficient for the largest particles (in contrast to the preceding case).

Transverse filtration, which occurs when the average flow is perpendicular to a sufficiently fine filter grid, represents a limit case. Particles accumulate on top of the filter; over the course of time, they form a filter cake, whose permeability diminishes as its thickness increases (figure 11.10b). We can liken the process to what we have observed in the context of sedimentation. Sediments form from a deposit of particles that is denser than the liquid containing them. In the case of a filter, however, a liquid is pushing particles along, which leads to more compact flows. Because the cake thickens over time, the filter must be regenerated regularly—for instance, by inverting flow direction. In industrial processes, an operation of this kind requires that filtering be stopped every now and then so that maintenance may be conducted, which is costly.

As a result, methods of *tangential (or cross-flow) filtration* have been developed (figure 11.10c). Liquid reaches the filter tangentially and emerges on the other side of this porous medium in a purified form. The liquid crosses the filter locally and creates a cake on the pores. In the case of these last two modes of filtration, the cake—more than the filter on its own—does the work. Here, however, the liquid sweeps the cake tangentially and erodes it constantly, replacing its surface with new grains after an equilibrium thickness has been reached. One industrial application is beer clarification. Tangential flow creates structures upstream from the flow that are looser and more deformable than those in cross-flow filtration. Structures of the same kind are also found in sand beds where flow occurs, which makes this problem similar to important challenges civil engineers face in coastal environments.

We have presented a simplified account of filtration, an essential process in engineering. Various forces are present together, especially the fluid flow accompanying the motion of grains, which has to be evaluated. The next, and final, chapter will turn to grains swept along by a fluid—a matter we have started to discuss. Chapter 12 will pay greater attention to the interaction between grain displacements and fluid motion.

12

GRAINS IN A FLUID

Panta rhei ("All things are flowing").
Heraclitus

This final chapter will offer the full "fluid perspective" and examine the behavior of more or less compact grain packings in a fluid at rest or in motion. A moving fluid contains a vast variety of immersed granular materials: sediments suspended in a river or a settling pond, sand blown by desert wind, underwater avalanches that can stretch for miles. Readers will call examples of their own to mind as we explore situations illustrating the essential role that fluid flow plays in the motion of grains. The subject of particles suspended in liquid or gas is so vast that it would merit a book of its own. The topic concerns fluid mechanics and the rheology of granular media in equal measure.

A LITTLE ORGANIZATION!

GRAIN CONCENTRATION

We'll start with the simplest example, of a unique solid sphere immersed in a flow field. Even when the concentration of particles is low, the flow induced by a grain in the fluid affects other grains. In other words, even though the particles may never touch each other, they nevertheless interact through the fluid. Here, the dominant phase is fluid, but within the grain-fluid mixture—which is known as a *suspension*—the effect of particles shows up in an increased viscosity of the suspension. In contrast, for high grain concentrations, such as pastes and muds, granular material represents the dominant phase; grain flow is governed by friction between them, as is the case for dry grains. This is why it's more accurate to speak of *immersed* granular matter.

Fluid modifies grain flow through viscous forces but also, sometimes, by inertia (as in the case of *debris flows*). By the same token, fluid can act as an agent of erosion, for instance in rockfill dams or at the bottom of a river. Conversely, it may play a lubricating role—as air does between the grains of a fine powder. Rheology, the science of flows of matter, concerns these materials, which combine grains and liquids.

GRAINS OF DIFFERENT SIZES

One can discuss the various flow regimes by taking the size and mass of grains as points of reference. Particles that are sufficiently small will remain suspended in the fluid under the effect of thermal agitation (described in chapter 1): the permanent restless motion of small objects as a manifestation of the thermodynamic temperature of a fluid. This constant motion grows in intensity the lower the mass of particles is. Thus, particles spontaneously disperse in space—a phenomenon described by the concepts of *molecular diffusion* and *Brownian motion*. Models can be devised where gravity plays no role when the liquid's density is adjusted to that of solid grains, and in cases where, in a fluid at rest and with very small particles, only molecular diffusion is at work. The density of the liquid can be close to that of the grains: for instance, the salinity of seawater allows us to float without sinking; salted water will also make a fresh egg bob on the surface.

As a rule, gravitation makes particles accumulate as they fall. It counteracts *Brownian diffusion*, which prompts them to disperse vertically. The result of this conflict is that particles form a layer of sediment at the bottom of the vessel where they are deposited. The higher the particles' mass is, the more quickly this layer forms. Thermal agitation accounts for small particles spreading in this manner. This is the realm of *colloids*, where physico-chemical interactions between grains predominate. Take a

Figure 12.1
The effect of debris flow in a river torrent in an Alpine mountain range. The lateral dams serve to stabilize the banks.

handful of soil from the yard and empty it into a transparent container filled with water. The large grains will quickly fall to the bottom; the smallest ones, after a certain time, will form dense layers on top of each other. The water that doesn't clear up entirely is full of tiny particles that remain in suspension.

And what about the largest grains? River torrents can carry blocks of considerable size in debris flows. Mixtures of water, mud, pebbles, and rocks created after heavy rain or in the accidental breaking of dams can cause considerable damage. The *lubrication** with a paste made of water, tiny grains, and clayey soils in general makes it possible for heavy rocks to slide over long distances.

THE NATURE OF FLUID
Of course, it's not the same thing to follow grains that fall in the air, water, or a viscous oil. We've named the property of matter—*viscosity*—that gives rise to a force opposed to the speed at which a small grain moves. If in a fall from one or two meters the speed of a ball increases continuously, it's because the friction effect of the air limiting the ball's velocity only proves significant at higher speeds, which are reached only in the course of a longer fall. For grains carried in the air—for instance,

the samaras of maple trees that turn like a helicopter blade as they slowly descend, or tufts of dandelions that call a parachute to mind—nature has found strategies for slowing descent by permitting a greater dispersion of seeds under the effect of wind.

Viscosity plays a major role for liquids. Drop a sphere into water—or, better yet, into water whose viscosity has been increased by adding cane syrup—and it will accelerate until friction, which increases with velocity, balances the weight reduced by the effect of buoyancy. For liquids of high viscosity like oil or honey, as well as for objects that are small enough, so-called *laminar flows** around a settling particle, described in chapter 11, are regular and predictable. In contrast, if the object in question is sufficiently large and the motion fast enough—say, a sphere one millimeter in diameter falling into water—the fluid flow induced by the fall will present eddies and whirling structures that are unpredictable: this is the domain of *turbulent flow*.* Recent advances account for the complexity of these flows and their modification, induced by particles whose motion is coupled to that of the fluid.

A SPHERE FALLS

THE STOKES SOLUTION
The motion of a single grain in a fluid provides a necessary precondition for studying suspensions controlled by the effect of viscosity. Yet again, our trusty sphere will lead the way as we watch it fall in a liquid of lower density (figure 12.2). The falling speed V under the effect of weight

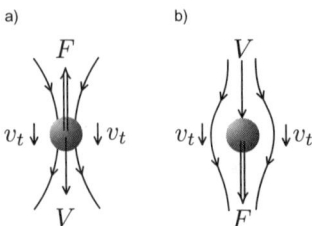

Figure 12.2
a) A sphere falling at a low Reynolds number in a liquid at rest. The force of viscous friction F is proportional to the sphere's velocity V. Liquid around the sphere is drawn along at velocity v_t, which decreases with distance from the sphere. b) A fixed sphere placed in a uniform flow of velocity V experiences a thrust force F.

corrected for buoyancy is vertical. The sphere induces motion in the fluid around and above itself; at the same time, it is slowed down by a frictional force F due to the fluid's viscosity, which increases proportionally to the velocity V. The *Stokes formula* expresses the relation between the velocity V of a sphere of radius R falling in a liquid of viscosity η and the force F (taking into account the Archimedes buoyancy force).

The Stokes Law

The Stokes formula expresses the proportionality between the speed v of a particle with radius R as it falls in a liquid of viscosity η, and the force of friction $F = -6\pi\eta RV$. This formula appears to be simple, but its demonstration is complicated—and not necessary for our purposes here. The minus sign indicates that the force is in the direction opposite to velocity: friction counteracts motion. The proportionality of F to V is a key result arising from the linear properties of viscous liquids, so-called *Newtonian fluids*: "Double the force, and you'll double the velocity." The proportionality of force to viscosity η is also a characteristic of viscous fluids. The factor R derives from the fact that the bigger an object is, the more resistance it will offer to flow.

Stokes's law is "robust": if one doesn't take the numerical coefficient of proportionality into account, it is the same for objects that have different shapes but the same average size. For a thin rod of length R or a flat disc of diameter R, the frictional force at a given speed will be close—up to a factor of two or three—to that of a sphere of radius R. This fact may seem paradoxical, but there's an explanation: viscous flows have a long range. If liquid motion is induced at points far from R, the whole liquid volume over a distance of the order of R is set in motion. Despite the complexity of shapes of all grain types suspended in fluid, at least this represents a simple result. There's no need to know the exact shape in order to evaluate the order of magnitude of hydrodynamic forces in a dilute suspension.

FLUID INERTIA

Drop a sufficiently large sphere into a big vessel filled with liquid, and you'll see that the sphere oscillates as it falls. This occurs in reaction to swirling motions of the fluid behind the sphere that were not present in the case of laminar flow previously discussed. A *turbulent regime* now prevails. The friction force F is given by $\rho C_X S V^2$. Force grows as the square of velocity—that is, faster than in a viscous liquid. It no longer depends on viscosity but on the liquid's density ρ: what counts is the *inertia* of the liquid brought about by the sphere's descent. S is the cross section of the obstacle seen by flow. The coefficient C_X takes into account the effect of the shape; it is small for elongated shapes with an unimposing profile (like the compact car manufactured by Citroën called "CX").

The Reynolds Number

The two different formulas for force F correspond to two different mechanisms. To appreciate their relative influence, it is possible to express their ratio: $\rho S V^2 / (\eta R V) = \rho R V / \eta$ because they represent quantities of the same nature (in the spirit of dimensional analysis, multiplicative numerical factors have been discarded here). This dimensionless quantity, or Reynolds number Re, plays a considerable role in fluid mechanics, enabling us to distinguish between "fast" and "slow" flows and to compare flows around obstacles under different conditions of velocity, viscosity, and size.

What matters is not any one of these factors on its own, but their dimensionless combination, which has a global significance. It makes little sense to say that a child is short or tall; it is meaningful, however, to say that a child is tall with respect to average size for his or her age. This analogic relation plays a key role in how physicists think. In fact, terms such as "rapid flow" and "viscous liquid" should be banned from scientific language, and only expressions such as large (or small) Reynolds number retained. A low Re value means that viscous forces dominate; this is the case for laminar flows; it's also the case for the Poiseuille flow in a tube illustrated in figure 11.3a. A large value means that inertia prevails and turbulent structures appear.

DIFFUSION AND MOBILITY

We have often referred to disordered motion displayed by "small" grains of matter as the result of thermal agitation. The significance of such movements can be evaluated in the context of sedimentation. Orders of magnitude are easy to remember. A sphere 0.5 μm in diameter and twice as dense as water falls at a speed of 0.5 μm per second. Its velocity due to thermal agitation at ambient temperature is also 0.5 μm per second. On this scale, the movements imposed by external forces and thermal agitation are comparable. The Reynolds number is of the order of one millionth: inertia effects are negligible!

A dimensionless number enabling us to predict whether a particle will display significant Brownian motion under the effects of flow is the *Péclet number* (Pe).* It replaces the Reynolds number for a flow: the molecular *diffusion coefficient* replaces the *viscosity*. The Péclet number compares the effects of flow and molecular diffusion. If it is large, disordered motion due to Brownian collisions proves insignificant relative to flow effects. In the opposite case, thermal agitation prevails.

INTERACTING PARTICLES

A SPHERE IN A TUBE

Let's start with the behavior at low Reynolds numbers. What happens when a single particle falls in a tube? The lateral walls reduce the flow induced by the sphere's motion and slow it down: if the diameter of the cylinder is ten times that of the sphere, the decrease of its velocity will be of the order of 10%. It's a surprising result: one might think that a sphere would be perfectly at ease in a tube ten times larger! This proves that, at a low Reynolds number, the effects of flow around an object make themselves felt at large distance from the object. Now consider a sphere approaching the bottom of the cylindrical vessel. It's difficult to get rid of the thin film of liquid between the sphere and the plane because of dominant viscous effects. Lubrication is at work; as noted earlier, it refers to friction between two solid bodies separated by a thin fluid layer. The effects bear on many problems in industry far beyond the scope of the suspensions discussed here and belong to the field of *tribology*. We have seen the importance of tiny surface irregularities at the interface between grains in the context of dry contacts. In a similar way, the thin liquid film between two objects that are extremely close and approaching each other limits their speed of approach.

INTERACTION BETWEEN TWO SPHERES

When a particle moves, the flow affects its neighbors. The same phenomenon holds for a flock of migrating birds or a pack of cyclists in a race. Beneficial interactions between the leader of the flock or pack and the others behind them are not symmetrical (the latter don't help the former!); this is characteristic of the high Reynolds numbers. In contrast, low Reynolds numbers entail reciprocal interaction between particles.

A BUNDLE OF SPHERES FALLING

If one places several spheres side by side in a fluid, the effect of their fluid-mediated interactions will prompt each one of them to fall faster than if it were alone: the bundle moves faster than a single sphere, as if the particles were attached (figure 12.3a). That might seem reasonable, but it is not the case in practice because of the geometry of the container: If the number of particles increases indefinitely, their falling speed—that is, the *sedimentation rate*—should become extremely high. In practice, however, the velocity for a large ensemble of particles decreases when their density reaches a high enough level!

The reason is the back flow of liquid that the particles have set into motion in the presence of a solid bottom surface of the container, which stops the downward fluid flow. Thus, in a closed tube, the counterflow velocity v_f slows down the fall of grains (figure 12.3b) and the sedimentation velocity decreases as its concentration grows.

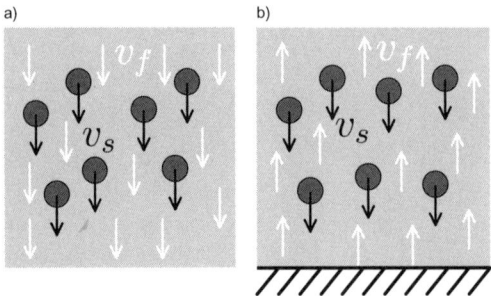

Figure 12.3
a) When a pack of grains sediments, the fluid is pulled downward. Its proper falling speed v_s is compounded by an additional component v_f induced by the fall of neighboring grains. b) If the grains are in a closed tube, an upward counter-flow v_f will take place between grains and slow them down; the settling rate diminishes as concentration increases.

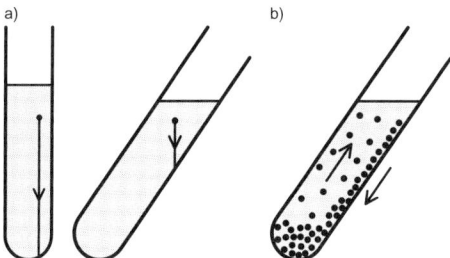

Figure 12.4

The Boycott effect: a) the length of a settling grain's trajectory in a vertical tube is greatly shortened in b), when the tube is tilted. The fall of particles and rise of fluid take place in different parts of the tube, leading to much faster sedimentation.

This illustrates the *Boycott effect*, named in recognition of the doctor who observed that when he tilted the tube in which he was measuring the sedimentation of red globules, the process occurred more rapidly than when the tube remained vertical. The particles descend along the low side of the tube, while the liquid rises along its upper side; the back flow has been channeled and has no effect on the fall of particles (figure 12.4)! The same effect is sought in order to improve foot traffic, when many pedestrians are moving in opposite directions.

SEDIMENTATION

Pour a dilute collection of microscopic grains into a test tube filled with a liquid (figure 12.5). Shake it to homogenize the suspension; then let the grains sediment. The upper part will clarify. The boundary between the clear liquid above it and the homogeneous suspension below falls at velocity v_s, which is also the speed at which the individual particles fall throughout the volume. When the initial concentration rises, this velocity declines as a result of viscous friction, as previously discussed; this decrease becomes more and more pronounced as the maximum concentration of spheres (that is, of spheres in contact) is approached.

Particles accumulate at the bottom of the vessel and form a loose pile of grains, a sediment whose thickness grows at a lower speed v_1, of the opposite sign, as particles get collected.

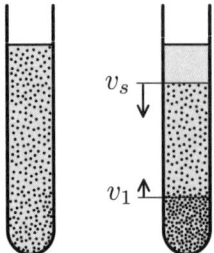

Figure 12.5
In sedimentation starting from a homogeneous suspension at left, identical particles will fall in liquid at an individual velocity v_s, which is also the velocity of the front holding pure liquid above. In contrast, the upper front of sediment rises at a lower speed v_1.

SEDIMENT FORMATION

In geological time scales, this process led layers of grains to form sedimentary rocks, which today cover three-quarters of the Earth's surface. Such rocks bear traces of strata produced by sedimentation and by the effects of flows. Once dried, sediment is generally not very compact; you can confirm as much when walking on the surface of a dried-out lake. Indeed, during deposit formation, the deceleration of particles as a result of the liquid's viscosity and lubrication effects between grains prevents them from reorganizing themselves enough to reach a high packing fraction.

CHEMICAL INTERACTIONS

The packing fraction can be even lower if chemical interactions occur between particles. This is the case for colloidal particles in which tiny grains attach to each other to form a fractal object that calls a snowflake to mind (figure 2.9). In the course of sedimentation, colloidal particles—for instance, those made from clayey mud—collide as a result of thermal agitation and stick to each other. This yields flakes whose settling rate grows with size. The process accelerates until these objects "full of void" come into contact, at which point they form a very weak random continuous lattice—a gel. If a 5% concentration of bentonite clay is placed in water, sedimentation stops when just half the suspension has clarified, but this 5% fills up half the volume!

The importance of physical chemistry can be gauged by observing the viscosity of a suspension of small silica grains; their modest size is responsible for strong attractions between particles. These grains tend to aggregate, which is countered by electrical charges of the same sign at the surface of particles. Varying the acidity—pH—of the solution makes these charges change. Electrophoresis allows this effect to be measured. The mobility of particles is charted as an electrical field is applied. In an acid medium, the silica particles move in the direction of the field, whereas they move in the opposite direction in a basic medium. Between these extremes, an *isoelectric* point exists, where the effect of charges disappears. The viscosity of suspensions reaches a maximum around this isoelectric pH value because particles no longer repel each other; they form large bundles, thereby increasing viscosity.

Electric charge effects are not the only cause of repulsion that opposes particle aggregation. The science and engineering of colloids is based on treatments for giving grains a surface layer that keeps them separate. Research on polymers has shown how to cover a solid surface with polymer chains. By way of comparison, consider opposite types of hair— tightly curled and perfectly straight, for example—to get an idea of how these microscopic polymers keep grains apart (for instance, the carbon grains suspended in India ink).

AGITATED PARTICLES

We have seen particles falling under the effect of their own weight in a liquid at rest and interacting with each other through a flow their motion has induced. How do collections of particles behave in the absence of sedimentation?

THE DANCE OF THE PARTICLES

We will start with a situation that permits us to disregard the weight of grains by adjusting the density of the fluid in which they are swimming to that of the grains. Conducting a free space experiment at zero gravity would also work, but that would prove much more costly!

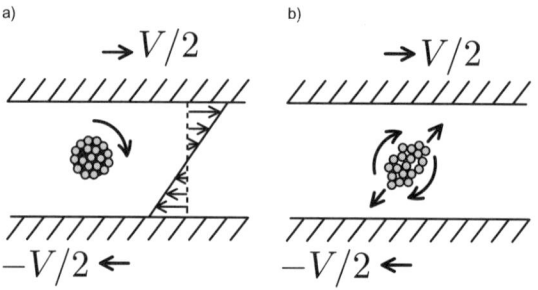

Figure 12.6
a) Placed between two parallel plates moving at opposite speeds, an agglomerate of consolidated spheres rotates at an angular velocity corresponding to the shear rate. b) If the spheres of this agglomerate are free to separate, the bundle stretches out at a 45-degree angle to the flow.

Simple shear, which we discussed in chapter 4 (figure 4.6), represents the most classic kind of flow. We'll start by putting a single sphere halfway between two plates moving at opposite speeds and separated by the liquid (figure 12.6). The liquid's velocity in the median plane is zero: instead of moving, the sphere rotates at an angular velocity γ, the shear rate (velocity divided by the distance between plates). This rotation is due to the friction of the liquid, which heads in opposite directions at the top and bottom of the sphere.

In a simple shear flow, this *rotation* effect is accompanied by an elongation effect: the liquid near the top is drawn to the right, and the liquid near the bottom to the left; in a sense, the liquid is being stretched. Other means exist for obtaining this elongation effect. For example, one can let a liquid flow through a small opening made on the bottom of a receptacle; elongation occurs because of convergence of fluid above the hole.

Let's consider an array of consolidated spheres in a general flow field. The spheres will turn around their own axes due to the rotational part of the flow field. But in response to the elongation part of the flow, they will change partners by attracting new grains and losing old ones. Here we have the elements of a random ballet whose choreographer is the nature of the flow.

FLUIDIZATION

Instead of grains falling in a liquid at rest under the effect of their own weight, imagine particles being held up from below by a jet of liquid or gas—say, a ping-pong ball floating at the top of a spout of water. If the friction the liquid exerts on particles balances the effect of their weight, they will remain suspended. If the speed of the fluid heading upward is high enough, the spheres' motion will be agitated by turbulent flow. This phenomenon is at the origin of *fluidization*,* which occurs with flow at a greater speed than with sedimentation.

Figure 12.7 shows a model experiment. An ensemble of identical grains has been placed on top of a grid that is fine enough to hold them, and a fluid is injected from below. As long as upward velocity remains low, the grains behave like a filter set in place, and the fluid passes through it (figure 12.7a). Above a velocity threshold at which the fluid's thrust balances out the weight of a particle, the bed of grains rises uniformly (figure 12.7b). A stable bed can be obtained at a value just above the fluidization threshold, but this does not prevent the grains from being in a state of constant movement with respect to each other. If the flow velocity is raised again, the particles' motions will grow erratic. The flow itself becomes turbulent, and bubbles of pure fluid rise through the bed so that two phases are present simultaneously—pure fluid, on one hand, and fluid containing grains, on the other hand (figure 12.7c). Pure fluid is less dense than fluid charged with particles, and the bubbles call to mind a convective phenomenon whereby a liquid that is hot and less dense than the averaage temperature rises up in "thermal plumes." Such instabilities prove quite important if fluidization involves a gas; because of the large difference in density between solid grains and a fluid, they can impair the functioning of industrial equipment that requires a high level of homogeneity in mixtures.

The fluidized bed may be deemed a new state of matter. Without a form of its own, it is not very dense but consistent. It has many applications in industrial engineering, where the goal is to maintain fine grains of matter in active contact with a fluid. This may involve, for instance, blowing hot air upward to make floating carbon powder react with it—a process that promotes efficient combustion. Indirectly, the effects of atmospheric pollution caused by incomplete combustion are reduced.

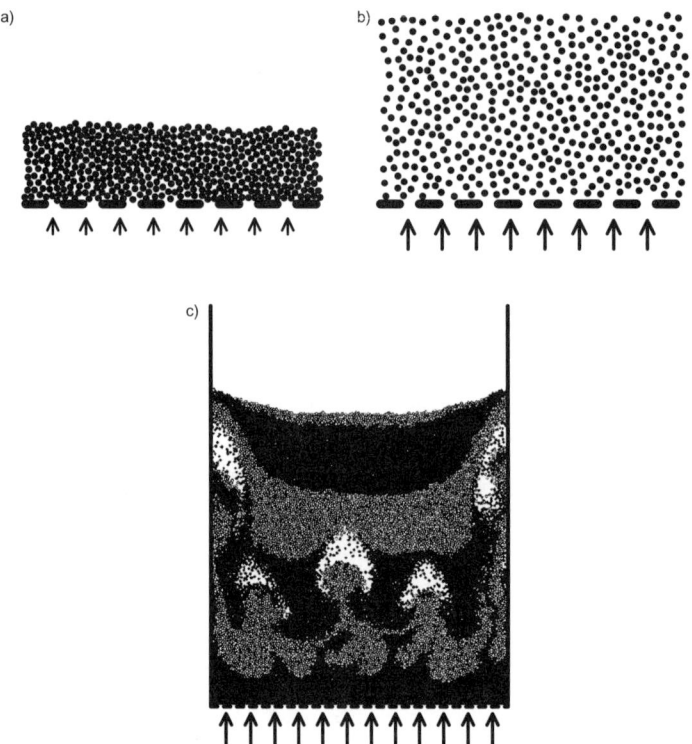

Figure 12.7
a) Subjected to an upward fluid flow, if the velocity of the rising fluid is insufficient, the fluid circulates through the porous space between the fixed spheres. At the fluidization threshold speed v_F, which is larger and opposite to the speed of sedimentation, the bed rises up since b) the weight of the spheres is overbalanced by friction. c) For higher fluidization velocities, the fluidized bed is unstable because of bubbles of fluid rising through it.

THE VISCOSITY OF A SUSPENSION

HOW VISCOSITY VARIES WITH CONCENTRATION

A simple method for evaluating viscosity η of a pure liquid is to measure the time it takes for a small sphere (at low Re number) to fall in the liquid: this is the principle at work in the so-called falling ball viscometer. Other devices are based on the principle of simple shear, for example the Couette viscometer, which consists of one cylinder rotating inside another fixed cylinder on the same axis. Laminar flows exhibit proportionality

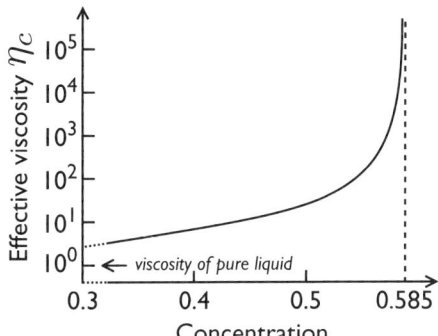

Figure 12.8
The effective viscosity of a suspension as a function of concentration C_c.

between the angular velocity and the applied torque; the suspension's effective viscosity η_s represents the ratio between these two quantities.

Figure 12.8, which shows how this quantity increases with the concentration (or packing fraction) of particles C_c, will guide our dicussion. Viscosity starts from $\eta_s(0) = \eta$—the value for the pure liquid—and increases in a linear fashion because of the viscous friction between grains induced by shearing. This variation becomes more rapid as particle concentration increases and grains get closer. Now, however, the mechanical effects between grains in contact contribute to the process. So long as conventional liquids are employed, and dilute suspensions, the measurement of viscosity yields the same value, no matter what kind of viscometer one uses. This is not the case at high concentrations because the arrangement of grains and, in consequence, energy dissipated as they move about depend on the type of flow at work.

A DILUTE SUSPENSION

Take a suspension with a small percentage of grains. Viscosity will increase because of friction between the fluid and particles. The result is summed up by a formula Einstein developed during his classic work on Brownian motion. The viscosity of a dilute suspension of particles of concentration C_c equals that of the suspending liquid η multiplied by a coefficient higher than the unity depending linearly on concentration: $\eta_s = \eta(1 + 2.5C_c)$; thus, for a concentration of 4%, viscosity will be higher than that of the pure liquid by 10%. One way of increasing viscosity, then, is to replace pure liquid with a liquid containing solid particles. The remarkable thing about the formula is that the viscosity doesn't depend on the size or shape of the particles.

AT HIGH CONCENTRATIONS

Viscosity increases faster and faster along with the concentration of particles. Eventually, the particles begin to collide and form an increasingly dense network. Forces are transmitted both by the contact network between particles (as in the case of dry granular flow), and by the fluid. Consider a pressure-controlled system like the one in chapter 8 (figure 8.1).

To study suspension shear flow under controlled pressure, a new *rheometer* was developed at Aix-Marseille University (figure 12.9). In this cylindrical apparatus, the granular packing is immersed in a liquid and confined by a porous cover that can let the liquid pass, and over which a constant confining pressure is applied. Since the liquid can easily pass through, the grains sustain all the pressure. The wall's rotation around the axis directly shears particles in a continuous fashion. In this way, the rheometer allows researchers to examine the grain assembly's resistance to shear in the presence of a liquid for different levels of confining pressure and fluid viscosity.

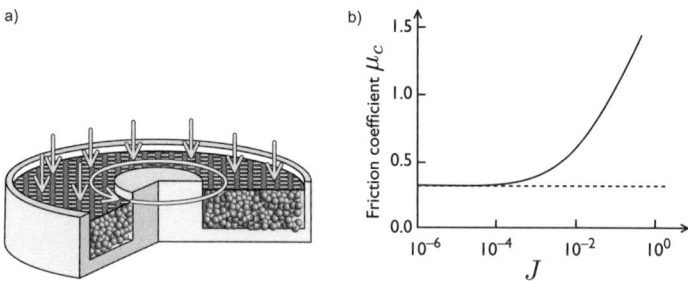

Figure 12.9

a) Rheometer for suspensions at controlled confining pressure: grains immersed in a liquid are confined between two fixed cylinders with a porous cover that lets through liquid (but not grains) as it rotates. Pressure is applied on the cover, and the suspension is sheared by the rotation of the inner cylinder. b) The suspension's coefficient of friction μ_c is measured by dividing the shear stress by the pressure applied to the cover. This curve shows how friction varies as a function of the parameter J, which is the ratio between the viscous stress the fluid exerts on grains and the confining stress.

Viscous Number

We have seen (see figure 8.1) that the viscous stress is proportional to the product of the liquid's viscosity η and shear rate γ. The ratio between the viscous stress and normal stress (pressure) P is denoted $J = \eta\gamma / P$. At the scale of individual particles, this dimensionless number, the *viscous number*,* measures the relative values of viscous force (given by viscous stress $\eta\gamma$ times the section R^2 of the particle) with respect to that due to confining pressure (given by confining pressure P times R^2. Note, in passing, that in a granular medium without applied pressure, the force induced by pressure at the free surface is simply the weight mg of a single grain. Since mass m varies as the cube R^3 of a particle's radius, J varies as $1/R$ —and therefore becomes smaller as grains increase in size.

The rheology of granular flow is characterized by the way the *effective friction coefficient* μ_c and the concentration C_c vary as a function of the *viscous number* J. The effect of J can be studied from two different points of view:

- From the *particle viewpoint*, the situation corresponds to that of dry flows. In this case, however, the rheology is governed by the *viscous number* J which accounts for the fluid viscosity—instead of the *inertial number* I discussed in chapter 8. When J increases (with a decrease of P, an increase of γ, or decreasing grain size), friction μ_c increases and packing fraction C_c decreases, as shown on figure 12.9b. The dense limit (large value of C_c) corresponds to low values of J, that is, to large values of P or low viscosity values. This is the quasi-static limit we discussed in chapter 8.

- From the *liquid viewpoint*, the suspension flow is characterized by effective viscosity η_c instead of effective friction coefficient μ_c. To measure η_c, the distance between mobile plates is fixed, so that the cell volume is constant. In other words, the volume is monitored instead of the confining pressure.

Although the effective viscosity is measured at fixed volume, its value is related to the effective friction coefficient measured at fixed confining pressure. Indeed, given the expressions of the shear stress $\tau = \eta_c\gamma = \mu_c P$ in terms of effective viscosity η_c and normal pressure P, and the definition of *viscous number* $J = \eta\gamma / P$, it is easy to see that $\eta_c = \mu_c\eta / J$. This relation allows one to obtain the evolution of η_c with J from the evolution of μ_c with J. It implies that the effective viscosity diverges as J tends to zero while at the same time the packing fraction tends to its critical value C_c.

As figure 12.8 shows, the rapid increase of effective viscosity and its divergence reflects the transition from a state dominated by the fluid (large values of J) to a state dominated by contacts between particles (low values of J). At this latter limit, there is enough jamming that particles cannot move with respect to each other, and they form a solid; the packing gets blocked when a constant volume is imposed on the granular assembly, and its effective viscosity diverges to infinity.

COMBINED EFFECTS OF VISCOSITY AND INERTIA

The foregoing analysis omitted inertial effects of both grains and fluid. This corresponds to a *viscous* regime. When the inertial effects of grains are significant—as is the case for rapid geological flows—the inertial forces of particles must be considered; the inertial number I and the viscous number J are to be taken into account simultaneously. Numerical simulations at the University of Montpellier have shown that inertial and viscous effects are additive: an inertial flow of dry grains on a slope, for example, resembles a viscous flow of grains immersed in a fluid. To understand this connection better, we should remember that the viscosity of a pure fluid results from the random motion of molecules. When agitated grains are added to a fluid, this new factor keeps on increasing the overall viscosity. However, there's an important difference between the two situations: the erratic motion of molecules is controlled by temperature, whereas the erratic motion of solid grains is governed by shearing and confining pressure (or the weight of grains).

FLUID INERTIA

For all that, granular inertia by itself does not mean that fluid is also in an inertial regime. The latter characteristic is evaluated by means of the Reynolds number, which must be considered alongside the values of I and J to know whether a flow is naturally turbulent. And there we said it ... *turbulence*, the name for one of the most formidable problems facing physics today! Great advances combining numerical simulation, high-power computers, and experimentation have enabled subtle measurements of turbulent fluctuating velocity fields, which have vastly improved our understanding of unstable flows. In particular, imaging technology following the trajectories of very tiny particles present in the flow (*particle image velocimetry*, or *PIV*) makes it possible to obtain a 3D map of a fluid's flow field. This approach has replaced invasive anemometric technologies.

The mixed flows of grains and fluid examined in this chapter have a low Reynolds number. A high Reynolds number is practically always encountered in the earth sciences: on large scales, the effects of a fluid's viscosity often prove to be secondary, compared with the effects of inertia.

The motion of small particles in a turbulent flow is present in a wide range of natural phenomena. Because this complex problem involves viscosity, inertia, and friction all at once, homogenization is often used: at an elementary level, particles held in suspension by flow turbulence and collisions between grains are then described in terms of effective viscosity, without a detailed reference to the local properties.

Sandstorms provide an impressive example. A sandstorm is characterized by a large cloud of dust advancing at the level of the soil, like a huge billowing wave. The cloud's movement derives from the difference in density between the air loaded with grains of sand and the less dense uncharged air in front of it. A gravity current of this kind can be simulated at home by pouring a thick drop of sugar syrup into a glass filled with water. It will spread out all over the bottom of the glass. The same mechanism is at work in underwater *turbidity currents*, where the flow induced by the sloping seafloor can be large enough to lift up immense rocks and cause a veritable underwater avalanche with dramatic consequences for nearby shores.

These last examples indicate that we have reached the end of the path we set out to travel in this book. At this point, the reins should be handed over to specialists in the various related fields of earth science and technology; ideally, our presentation of the principles of granular media will shine a fuller light on their studies. Looking back a few decades—and considering one of the authors' previous books on the subject and related works—it's clear that great advances have been charted since then, and many findings have made their way into a more unified field of scientific endeavor. We are confident, then, that shared efforts to arrive at more complete models of understanding these applications will bear even more fruit in years to come.

CONCLUSION

This book was inspired by the interest that granular materials have held since time immemorial, both in terms of practical uses and as a source for models developed in a host of research fields. We are convinced that granular materials will continue to inspire other domains of science, and that they offer the key to describing a vast range of phenomena—all of which may be interpreted by looking at a simple bag of marbles! Nobel laureate Richard Feynman asked his students at Caltech what best sums up our understanding of the world. His own answer: "Matter is made of atoms." He would go on to demonstrate everything atoms tell us about states of matter. The feature atoms share with grains is that, in forming granular packings according to their own laws of interaction, they give rise to a rich panoply of material properties; in either case, the common denominator is piling, stacking, packing, or arrangement.

Throughout this book, we have insisted on the importance of geo-metrical disorder in assemblies of grains. This is not a matter of a slight deviation from crystalline order that one finds in solid materials on an atomic scale. Instead, it concerns the "high disorder" of passengers piling into the subway train at rush hour. As we have seen, collective behaviors emerge within apparent chaos: force chains concentrate stresses onto a low number of grains, micro-arches form at exits and control the flow rate, and privileged contact directions lend an anisotropic character to granular media undergoing deformation. Static and dynamic properties arise from these internal structures, which enable granular materials to adapt to the stresses and deformations to which they are subjected. This "accommodating"—plastic and flexible—quality lies at the heart of many industrial applications. It is the reason for the "magic" that can be performed on the beach, with modeling clay, or with "kinetic" sand sold in toy stores; instead of stacking cubes, one organizes disorder, and a new form emerges.

Is it possible to suppress or weaken force chains in order to encourage grain flow? Can a granular edifice be reinforced to obtain lighter, biodegradable, more resistant, and tougher materials? How does one control the permeability of assemblies and their surfaces for purposes of filtration, or to catalyze new molecules? Can the connections between grains be adjusted to control thermal or electric conductivity, or even the transmission of waves? What's the way to encourage the self-assembly that produces such astonishing results among nanoparticles? These are all questions that motivate practitioners of what might be called "grain engineering." Just as a builder combines blocks, stones, and other elements to realize an architectural construction, we must choose or manufacture grains carefully so they yield an "archi-texture" that displays desired properties deriving from its natural disorder. This flight to conceive new metamaterials has only just begun.

The composition of granular materials is not expressed only in terms of chemical species. It's just as important to specify the range of sizes and shapes of grains, as well as the forces induced by solid or liquid phases, which modify interactions at the level of contacts. Methods and procedures for preparing grains change the way they pile up and assemble. In this vast range of possibilities, it's not easy to predict the combination of parameters that will yield the best results. Theoretical models, high-resolution imaging, and specific measurement tools help to determine strategies to create optimal structures for ideal textures. Three-dimensional printers enable us to manufacture grains of different shapes and sizes by adding layers progressively. Treating surfaces mechanically, chemically, or by means of plasma helps to control small-scale interactions. Systematically studying the properties of these grain assemblies, which represent the sites of constant improvements, will surely lead to textures with unheard-of properties in the near future.

Finally, we should remember that grains occupy a privileged position between two extremes: the scale of atomic interactions at one end, the scale of piles at the other. Sciences of molecular interactions that investigate the properties of materials constituting grains (physical chemistry) and those of aggregates that grains constitute (mechanics and physics) are usually studied independently. This book has sought to bring these two realms of research closer together. In the wonderful poem quoted at the beginning of chapter 1, William Blake declares that a grain of sand contains the secrets of the whole world. A single grain at human scale opens up to a double horizon: from the microscopic state of local interactions to the macroscopic organization of granular matter, it provides the background for new developments. A grain is a universe in the making.

ACKNOWLEDGMENTS

We dedicate this work to the memory of the recently departed Jean Jacques Moreau, professor of mathematics at Montpellier. The numerical tools he developed while researching nonsmooth mechanics (in particular, contact dynamics) opened a new field of virtual experimentation in granular materials. The authors of this book are indebted to his generosity and pedagogy.

We seize this opportunity to honor the memory of Robert Behringer (Duke University) as a prominent pioneer of modern research on granular matter who made a seminal contribution to the grain-scale description of granular flows.

We are grateful to Ken Kamrin (MIT) for his thoughtful foreword to this book. We also extend thanks to colleagues at Montpellier, Paris, and MIT. Informal day-to-day exchanges are how ideas are developed and tested, and information shared. Henri Van Damme (ESPCI ParisTech) and Jean Vaillancourt are warmly thanked for corrections and suggestions that were taken into account in this book. We thank Heinrich Jaeger (University of Chicago) for sharing with us his exciting research on aleatory structures and Roland Pellenq (MIT, CNRS) for sharing with us his vast knowledge of materials science.

We would also like to thank all our colleagues who offered advice or provided the documents included here.

GLOSSARY

Anisotropy. Accounts for effects that depend on direction in space.

Apollonian packing. Model packing comprising multiple layers of spheres, arranged in such a way that the voids between larger spheres are filled by smaller spheres.

Arching effect. Formation of vaults involving several grains that are able to support significant stress and relieve the effect of pressure on other grains.

Atomic force microscopy (AFM). By measuring the force between a very fine point moving perpendicular to a surface, it is possible to characterize the surface on a nanoscopic scale.

Brownian motion. Random behavior of particles sufficiently small to move as a result of the thermal agitation of molecules. Molecular diffusion is obtained when an ensemble of particles in suspension are spread by Brownian motion.

Capillarity. Effect resulting from the surface tension of a liquid; manifest in the surface forces of a liquid or its interface with a solid.

Ceramics. By firing clayey pastes and, more generally, nonmetallic granular materials, solid structures are obtained.

Clay. It is made of elementary particles of sizes smaller than a micron and composed of extremely fine platelets of aluminosilicates.

Coefficient of friction. When a shearing displacement is applied between two surfaces of materials in contact, there is a resisting force to motion parallel to the interface. It opposes the motion (static) or brakes it (dynamic). The ratio of this force to the normal force that pushes the two materials against each other is the coefficient of friction.

Coefficient of restitution. In contrast to a gas, where collisions between particles are elastic, only part of the kinetic energy is restored in a collision between two solid grains. The other fraction is dissipated in the form of heat.

Cohesion. Manifestation of permanent forces between grains of matter bonded together.

Colloidal crystals. Material made of monodisperse silica spheres of micrometric size forming a periodic network. In colloidal crystals, repulsive interactions between charged grains, which are generally of the same size, lead to periodic organizations in space on the scale of a few thousand angstroms.

Colloids. They are made of suspended particles sufficiently small for the effect of brownian motion to be significant. Colloids can be obtained by chemical means. The great variety of possible organizations reflects the electrical and physico-chemical interactions between particles.

Concrete. This word describes a dense assembly of granules of various sizes, a cementing material that consolidates them, and components that can reinforce mechanical properties, such as fibers. High performance concretes are obtained from assemblies of optimized size distributions, fibers, and chemical treatment.

Consolidation. Creation of cohesive bonds between solid grains.

Coordination number. The number of neighbors close to a site in a lattice. The average number of contacts between grains in a pile.

Coulomb's friction. At rest, it is the maximum force that an object pushed on a surface can exert along the surface. In the sliding motion of the object on the surface, the Coulomb friction is the force resisting motion.

Critical packing fraction. The value reached by the packing fraction of a granular material after a long shear deformation in a given direction.

Critical phenomenon. Behavior around a critical point where a change of state occurs—such as magnetism or phase transition. The properties of a body around this point do not depend on local details.

Critical state. At the threshold of a control parameter for a physical system (for instance, temperature), a phase transition occurs between different states. The behavior of material in the vicinity of the threshold possesses "universal" characteristics independent of the particular system. In granular flows, the critical state is reached after a long shear deformation. The packing fraction and internal friction angle are said to have their critical values in this state.

Deformation. Modification of relative positions of elements in a solid or liquid, or grains in a granular material, when stresses are applied.

Diffusivity. Characteristic coefficient of a physical system that measures the spontaneous spread of one material within another—for instance, a drop of ink in water or on blotting paper (see *Mobility*).

Dilatancy. A compact granular medium subjected to shearing increases in volume (reduces in packing fraction) to allow for relative motions between grains.

Dislocations. Defective position in a periodic arrangerment of atoms, responsible for large-scale loss of position order. Dislocations and accumulations of dislocations at grain boundaries underlie mechanical properties of typical metals and alloys.

Disorder. Randomness in particle positions.

Elasticity. Characteristic of a solid that relates elastic stress and deformation. In the case of small deformations, the stress is proportional to deformation (linear elasticity). At higher levels of deformation, nonlinear and irreversible behaviors are observed (plasticity, rupture). A small modulus of elasticity means that a material can be easily deformed.

Elongation. The extension of a solid subjected to a traction. It is generally accompanied by a reduction of size in the transverse direction.

Filtration. An ensemble of techniques used to separate grains according to geometrical criteria.

Fluidization. Technique in process engineering for keeping solid particles suspended by a rising vertical current of liquid or gas.

Force chains. In a confined packing of grains, the stress is carried by contact forces between grains. The grains submitted to strong forces form continuous long chains often propped up by much smaller forces.

Fractals. Geometrical objects and structures whose organization does not change, in detail or on average, when considered at different length scales. Fractal objects are characterized by a non-integer dimension D_F called *fractal dimension*.

Fracture. The final stage of deformation for a solid, triggered by the propagation of cracks leading to the split of the solid into several fragments. Brittle fracture occurs during a purely elastic deformation whereas ductile fracture takes place at the end of a plastic deformation phase.

Gel. A generally weak and easily deformable solid structure made of a three-dimensional molecular lattice in a solvent. The network can be composed of polymer chains. At the threshold of a critical concentration of bonds, transition occurs from a sol to a gel, with the appearance of elastic behavior.

Graphene. An extremely thin film made of a single or a few layers of graphite crystal.

Hard sphere model (or hard disc model in plane geometry). Model for interaction between particles that cannot overlap. Otherwise, we have a model of "soft spheres." Numerical models and analog simulations of hard (or, less commonly, soft) spheres form the basis for most of what we know about the physics of gasses, liquids ... and granular materials.

Hertzian contact. It describes the deformation of two elastic spheres pressed against each other as a function of applied force.

Homogenization. In low-disorder materials, certain large-scale properties may be determined by considering a representative sample of small volume containing a sufficient number of elements (atoms, grains) to exhibit well-defined properties independently of the size of the sample. These properties can then be expressed in terms of those of the elements.

Hydrostatic compression. The stress inside a material due to the applied external stresses. In a liquid, the hydrostatic stress is normally applied to any surface, and its value is the same in all directions. In a granular material, the stresses can be different in different directions.

Imbibition. One of the two phases (the second is drainage) of filling or emptying a porous medium by means of a fluid that wets pores in the presence of another, nonwetting fluid.

Inelastic collision. Collision between two particles with energy dissipation. The inelasticity of collision is characterized by a coefficient of restitution.

Inertial number. Dimensionless quantity comparing the inertial effects of grains in motion to effects induced by confining pressure. It plays a role in granular materials that is similar to the Reynolds number for fluid flows.

Laminar flows. May be achieved for sufficiently slow flows on a small scale or for highly viscous fluids. Laminar flows are characterized by a low Reynolds number. The flow rate is proportional to a pressure difference. If pressure conditions are independent of time, a steady flow occurs.

Laplace-Young law. The proportionality of pressure difference between two sides of a liquid interface to its mean curvature. The proportionality coefficient is its surface tension.

Lithostatic. In a soil, the weight of upper layers makes the vertical pressure increase with depth. The horizontal pressure increases with depth at a different rate. This is compared with hydrostatic pressure, which is the same in all directions at a given depth.

Long-range order. Characterizes an order or correlations between points that are distant from each other, as in a crystal lattice. Defects such as dislocations lead to the loss of long-range order.

Lubrication. Operation that consists of interposing a fluid film between two solids in order to reduce friction between them.

Mean curvature. It is given by the sum of the inverse principal radii of curvature (figure 9.3). It is involved in the capillary phenomena presented in this book (Lapalce-Young law).

Metamaterials. Arrangements of grains at a sub-micron scale; their properties bear on many recent applications, in particular 3D printing processes and the propagation of waves.

Microscopic (nanoscopic). Applies to systems with sizes of the order of one micrometer (one nanometer).

Mobility.   When subjected to a force, a particle will move in the direction imposed by this force F at a speed v that is proportional to it: $v = \alpha F$. The coefficient α is the particle's mobility. It describes the particle's propensity to move in a process of diffusion.

Nanoscopic (microscopic).   Applies to systems with sizes of the order of one nanometer (one micrometer).

Opals.   Naturally occurring close-packed lattices of hydrated silica. Since the characteristic distances of this crystal are of the order of the wavelength of visible light, selective reflections result in iridescence, which makes these jewels prized.

Packing fraction (C).   Fraction of volume occupied by grains in a packing.

Péclet number (Pe).   Number characterizing the relative efficiency of flow with respect to molecular diffusion effects in mass transport phenomena. At low values, molecular diffusion prevails.

Pedology.   Soil science (nothing to do with feet podology!).

Percolation.   Describes the progressive variation of connectivity and associated properties in a system with randomly distributed sites or particles. The percolation threshold corresponds to the percentage of contacts at which a continuous path appears for the first time between the particles. For a mixture of conductive and insulating grains that are geometrically identical, this threshold roughly corresponds to a 30% proportion of conductive spheres.

Permeability (κ), or hydraulic conductivity.   A characteristic property of a porous medium. Its practical unit is the *darcy*, equal to a squared micron. *Darcy's law* holds that the flow rate in a porous material due to pressure difference between two points is proportional to the permeability of the material.

Phase transition.   Most substances can exist in different physical states. A phase transition corresponds to passage from one state to another. For instance, water passes from solid state to liquid state when it reaches a critical temperature (this melting point is $0°$ C at atmospheric pressure).

Photoelasticity.   It permits the study of a solid transparent medium whose optical index is a function of the value and direction of the stress applied to it. Between crossed polarizers, the light transmitted across the medium reflects the inhomogeneity of the stress inside the material.

Plasticity.   Beyond a solid's limit of elastic behavior, deformation becomes irreversible, and the relation between stress and deformation becomes nonlinear.

Poiseuille flow.   The flow of a viscous fluid in a cylindrical tube with a low Reynolds number. The effect of viscosity at the walls leads to a flow rate that varies as the square of the cross section.

Porosity (ϕ).   Volume fraction of voids in a packing. It is complementary to packing fraction C: $\phi = 1 - C$.

Pressure. 1) *uniaxial*: the force per surface unit applied to a body in a given direction; 2) *hydrostatic*: pressure applied uniformly perpendicular to the surface of a whole body, as within a liquid; and 3) *lithostatic*: the pressure's vertical component that varies with the weight of the soil layer.

Random (or Brownian) walk. Spontaneous random motion of small particles as a manifestation of their absolute temperature ($T + 273$ degrees).

Representative elementary volume (REV). REV applies when, beyond a certain volume around a grain, average properties no longer depend on the volume at which they are evaluated.

Reynolds number (Re). Dimensionless combination of viscosity, size, and velocity in a flow for characterizing laminar and turbulent regimes. Another way to define the Reynolds number is to say that it measures the relative roles of convection and diffusion for velocity.

Rheology. The science of material flows. The famous dictum *panta rhei* ("All things are flowing") applies to the viscosity of liquids, the plasticity of a solid, and different modes of deformation for soft matter.

Sandstone. It is obtained over long periods of time by the deposit of ions transported by liquid circulating between grains of sand (*diagenesis*).

Saturated media (unsaturated media). Porous media filled entirely (or in part) by a fluid phase such as water.

Sedimentary rock. Material that results from bonding between grains over time (e.g., sandstone). See *Sandstone*.

Sedimentation. Deposit of grains at the bottom of a liquid layer in which they were suspended.

Shearing. Deformation of a solid or liquid such that the displacements differ in different directions. An example is a piece of material stretched in one direction and compressed in the perpendicular direction. Another example is the unequal parallel motions of two successive layers. A material that is compressed isotropically (equally in all directions) is subject to no shear.

Short-range order. Characterizes the local organization of elements (discs, spheres) around a central element. It generally exists for compact disordered packings of grains.

Simple shear. Deformation of a solid or liquid contained between two parallel plates that are subjected to a relative displacement. Two modes are possible, either maintaining distance between plates (in which case the normal pressure exerted by the flow on the plates can vary), or applying a confining pressure (in which case the distance between the plates or the volume of the material can vary slightly).

Sintered materials. Solid coherent materials obtained by *sintering* a granular packing under the effect of temperature and pressure.

Sintering. Operation for binding grains together to obtain a coherent material. Sintering can occur by adding foreign matter or, more frequently, under the effect of heat in a compact assembly of grains.

Soft matter. Describes the rheological properties of certain easily deformable materials. Polymers, gels, soap bubbles, suspensions, muds, and, by extension, granular media are examples of soft matter.

Sol. In the sol-gel phase transition, a macromolecular structure passes from a disconnected ensemble that keeps it in a fluid phase (sol) to a connected ensemble that has the properties of a solid (gel). Percolation theory has been applied to this transition.

Specific surface area. Total surface per unit volume in a discrete or porous material. Its unit is the inverse of a length. Its order of magnitude is sometimes close to the inverse of pore size.

Stereology. Discipline for determining properties of objects in space on the basis of their two-dimensional cross sections, provided that the system is randomly distributed and isotropic.

Stress. Force per unit area acting on a surface inside or at the boundary of a material. Pressure is a normal stress (perpendicular to the surface). Shear stress is the difference between stresses in different directions.

Surface tension (interfacial). Force per unit of length—or energy per unit area (which amounts to the same quantity)—accounting for the cohesion of the free surface of liquid (or an interface between two liquids or two solids).

Synchrotron (light). Very intense distribution of light with a large continuous distribution of wavelengths, emitted as a result of the curved path of particles in an accelerator.

Thermal agitation. The spontaneous and restless manifestation of thermodynamic temperature T_t in a diluted system of so-called Brownian particles. The temperature of a gas T is $T_t + 273$ *degrees.*

Tomography. Analysis of the mean bulk of geometrical properties in a heterogeneous material by means of images obtained from its cross sections. See *Stereology.*

Tribology. The science of contacts and friction. This classic domain has witnessed major developments recently, connecting microscopic and multi-scale research on solid surfaces and the physical chemistry of interfaces. In this sense, tribology may be classified among fields devoted to the study of soft matter.

Turbulent flow. At high Reynolds numbers, flows vary in unpredictable ways over time. Few flows of this kind have been presented in this book, because the grains and the gaps between them, through which a fluid circulates, are often weak.

Uniaxial compression. Effect of pressure forces exerted along a given direction.

Van der Waals force.   From the name of a Dutch physicist whose dissertation, written more than a century ago, provides the basis for our understanding of the physical states of pure substances. Van der Waals interaction characterizes the force between ensembles of atoms; it is electrical in origin. For two spherical grains, force grows proportionally to size and in inverse proportion to the distance between them. In practice, van der Waals interactions prove significant for grains less than a micron in size.

Viscosity.   Material parameter expressing the fluid's resistance to flow. This resistance results from internal forces of friction within the fluid, which are at the origin of dissipative processes and transform part of the mechanical energy available to it into heat.

Viscous number.   In a dense suspension of grains in a fluid, the viscous number is a dimensionless quantity measuring the relative importance of the average viscous force acting on the grains during flow with respect to the average weight of the grains (or forces between grains due to a confining pressure acting on the grains). For low values of the viscous number, the suspension behaves as a dry granular flow, whereas at high values it behaves like a dilute suspension flow.

Voronoï diagram.   Graph obtained by joining the centroids of particles in a granular assembly.

BIBLIOGRAPHY

Allègre, C. *From Stone to Star: A View of Modern Geology*. Harvard University Press, 1994.

Andreotti, B., Y. Forterre, and O. Pouliquen. *Granular Media: Between Fluid and Solid*. Cambridge University Press, 2013.

Bideau, D., E. Guyon, and J. P. Hulin. *La Matière en désordre*. EDP Sciences, 2014.

Bideau, D., and A. Hansen, eds. *Disorder and Granular Matter*. North Holland, 1993.

Cabane, B., and S. Hénon. *Liquides, solutions, dispersions, émulsions et gels*. Belin, 2007.

Coussot, P., H. Van Damme, and C. Ancey. "Des solides coulants." *Pour la science* 273 (2000): 34–40.

de Gennes, P.-G., F. Brochard, and D. Quéré. *Capillarity and Wetting Phenomena*. Springer, 2005.

Duran, J., and A. Reisenger. *Sand, Powders and Grains*. Springer, 2012.

Falcon, E., and B. Castaing. "Electrical Conductivity in Granular Media and Branly's Coherer." *American Journal of Physics* 73 (2005): 302.

Fontaine, L., and R. Anger. *Bâtir en terre*. Belin, 2009.

Gravish, N., and D. I. Goldman. "Entangled Granular Media." In *Fluids, Colloids and Soft Materials: An Introduction to Soft Matter Physics,* edited by A. Fernandez-Nieves and A. Manuel Puertas, 341–354. John Wiley & Sons, 2016.

Guinier, A. *X Ray Diffraction: In Crystals, Imperfect Crystals, and Amorphous Bodies*. Dover, 2003.

Guyon, E., J.-P. Hulin, L. Petit, and C. D. Mitescu. "Quasi-Parallel Flows—Lubrication Approximation." In *Physical Hydrodynamics*, 277–307. 2nd ed. Oxford University Press, 2015.

Guyon, E., and J.-P. Troadec. *Du sac de billes au tas de sable*. Odile Jacob, 1994.

Iverson, R. M. "The Physics of Debris Flows." *Review of Geophysics* 35 (1997): 245–296.

Jaeger, H. M., J. R. Nagel, and R. P. Behringer. "Granular Solids, Liquids and Gas." *Reviews of Modern Physics* 68 (1996): 1259—1273.

Kerisel, J. *Pierres et Hommes*. Presses Ponts et Chaussées, 2005.

Lapaire, J., and P. Miéville. *Le Sable et ses mystères*. BRGM éditions, 2012.

Mattauer, M. *Ce que disent les pierres*. Belin, 1998.

Powders and Grains Video Playlist. YouTube (2017). Contains forty-nine videos about granular matter.

Radjai, F. "La double vie du sable." *La Recherche* 304 (1997) 44–46.

Radjai, F., S. Nezamabadi, S. Luding, and J. Y. Delenne, eds. *Powders and Grains 2017—8TH International Conference on Micromechanics on Granular Media*. Open Access Proceedings. *EPJ Web of Conferences* 140 (2017). https://www.epj-conferences.org/articles/epjconf/abs/2017/09/contents/contents.html.

Sands, D. E. *Introduction to Crystallography*. Dover, 2003.

Schroeder, M. *Fractals, Chaos, Power Laws: Minutes from an Infinite Paradise*. Dover, 1991.

Van Damme, H. "Concrete Material Science: Past, Present, Future Innovations." *Cement and Concrete Research* 112 (2018): 5–24.

Van Damme, H. "La terre, un béton d'argile." *Pour la science* 423 (2013): 50–57.

Van Damme, H., and H. Houben. "Earth Concrete: Stabilization Revisited." *Cement and Concrete Research* 114 (2018): 90–102.

IMAGE CREDITS

Figures and illustrations are by the book's authors, with the exceptions listed below.

Figure 1.3. Photograph by Alfredo Taboada, Geosciences, University of Montpellier.

Figure 2.6. Photograph by Rémy Guadagnin, Jeunesse Préhistorique et Géologique de France (JPGF).

Figure 2.8. Micrography by Thierry Ruiz, Laboratoire IATE Montpellier, and Didier Cot, Institut Européen des Membranes, University de Montpellier.

Figure 2.9. Micrography by David Weitz, Harvard University.

Figure 2.10. Photo credit: D. E. Brownlee, University of Washington, and E. K. Jessberger, Institut für Planetologie, Münster.

Figure 3.1c. Photo credit: T. Wittermans.

Figure 3.2. Courtesy of Knut W. Urban, Institute of Solid State Research, Jülich.

Figure 3.8. Photo credit: A. Imhof, Debye Institute, Utrecht University.

Figure 3.9. Photograph by G. Meille, *Sac de billes*, Odile Jacob.

Figure 3.11. Courtesy of Jérome Lagoute, Laboratoire Matériaux et Phénomènes Quantiques, Université Paris-Diderot and CNRS.

Figure 3.14b. Image made with Neper software by Romain Quey, CNRS, École des Mines de Saint-Étienne.

Figure 3.15. Courtesy of D. V. Stäger and H. J. Herrmann, ETH Zurich.

Figure 4.1. Courtesy of M. Cloitre and C. Allain, ESPCI ParisTech.

Figure 4.2. Photo credit: Karola Dierichs, ICD Aggregate Pavilion 2015, Institute for Computational Design, University of Stuttgart.

Figure 4.3. Courtesy of P. Chaikin, Princeton University.

Figure 4.5. Simulation by E. Azéma, Laboratoire de Mécanique et Génie Civil, Université de Montpellier.

Figure 4.9. Rights reserved. Document provided by R. O. Pouliquen, CNRS, Institut Universitaire des Systèmes Thermiques Industriels, Université d'Aix-Marseille.

Figure 5.3. From I. Hutchings, "Leonardo da Vinci's Studies of Friction," *Wear* 360–361 (2016): 51–66.

Figure 5.7. Courtesy of J. Rajchenbach, Laboratoire de Physique de la Matière Condensée, Université de Nice-Sophia-Antipolis.

Figure 7.1. Photo credit: R. Behringer, Duke University.

Figure 7.2a. Simulation by C. Voivret, Laboratoire de Mécanique et Génie Civil, Université de Montpellier.

Figure 7.2b. Simulation by D. H. Nguyen, Laboratoire de Mécanique et Génie Civil, Université de Montpellier.

Figure 7.6. Simulation by J.-J. Moreau, Laboratoire de Mécanique et Génie Civil, Université de Montpellier.

Figure 7.7. Photo credit: L. Vanel and É. Clément, Laboratoire de Physique et Mécanique des Milieux Hétérogènes, ESPCI ParisTech.

Figure 8.3. Photo credit: R. Behringer, Duke University.

Figure 8.6. Simulations by D. Rakotonirina, Institut Français du Pétrole et Énergies Nouvelles and A. Wachs, University of British Columbia, Vancouver.

Figure 9.10. Courtesy of Thierry Ruiz.

Figure 9.11. Photo credit: R. Anger and L. Fontaine.

Figure 9.12. Photo credit: R. Zwart, Rotterdam.

Figure 9.13. Photo credit: J. Barès, Laboratoire de Mécanique et Génie Civil, Université de Montpellier.

Figure 9.14. Courtesy of Henri Van Damme.

Figure 10.2b. Photo credit: L. Olmos and D. Bouvard, Institut National Polytechnique de Grenoble.

Figure 10.3. Photo credit: L. Olmos and D. Bouvard, Institut National Polytechnique de Grenoble.

Figure 10.5. Micrography by M. Regourd and H. Van Damme, ESPCI ParisTech.

Figure 10.6. Micrography by E. Chichti, Laboratoire IATE Montpellier.

Figure 10.8. Simulation by V. Topin, Laboratoire de Mécanique et Génie Civil, Université de Montpellier.

Figure 11.2. Figure adapted from E. Guyon and J.-P. Troadec. *Du sac de billes au tas de sable*. Odile Jacob, 1994.

Figure 11.6. Document provided by J.-P. Hulin.

Figure 11.7. Simulations by N. S. Martys, National Institute of Standards and Technology, Gaithersburg, MD.

Figure 11.9. Document provided P. A. Scholle, American Association of Petroleum Geologists.

Figure 12.1. Courtesy of S. Gominet, Institut des Risques Majeurs, Grenoble.

Figure 12.7b. Simulation by S. McNamara, Institut de Physique de Rennes, Université de Rennes.

INDEX